CW00842495

FUR SEAL ISLAND

*By the same author*

THE YEAR OF THE TAWNY OWL
THE YEAR OF THE WEASEL

# FUR SEAL ISLAND

*An Environment in Peril*

Paul Thomas

SOUVENIR PRESS

First published 1990 by Souvenir Press Ltd,
43 Great Russell Street, London WC1B 3PA
and simultaneously in Canada



ISBN 0 285 62969 7

Printed and bound by WBC Ltd, Bristol and Maesteg

# CONTENTS

# ACKNOWLEDGEMENTS

My sincere thanks to the following people who have helped with the making of this book:

Roger Gentry of the National Marine Fisheries Service, National Marine Mammal Laboratory in Seattle, Washington State, USA, who supplied much information concerning the attendance behaviour of the fur seal.

Bernard Walton and Richard Brock of the BBC who arranged for me to visit the breeding colonies at St George, and to Bernard in particular for his encouragement in the creation of the story.

Hilary Westmacott for reading the manuscript and providing information concerning parasites.

# PROLOGUE

This is a saga of a brooding, misty island set in a cold northern sea, and of a determined, solitary creature fighting her way back, year after year, to breed on a bleak and crowded shore—a shore where man, over the last two hundred years, has slaughtered millions of her kind and is now jeopardising the survival of her species by his over-exploitation of her ocean.

Something is drastically wrong in the North Pacific and it is in this ocean that the northern fur seal lives. A gross imbalance exists, which threatens the very life of the oceans. Among the nine fur seal species of the world, only the northern fur seal has been found to have suffered a recent decline, despite the fact that man no longer kills fur seals for commercial purposes. Scientists are only now beginning to piece together the full implications of the tragedy and as yet have no clear idea why marine life should be declining. A combination of factors is to blame, not least the so-called 'factory' fishing methods which are currently employed indiscriminately over all the high seas and in particular across important animal migration routes.

The oceans of the world are the common heritage of us all. They are the last great sanctuary where large wild populations of invertebrates, fish, mammals and birds can live, feed and breed in their natural environment. The story of the northern fur seal gives us some insight into ocean life in general and is a specific example of the detrimental effect of human interference, an interference which has been monitored and recorded in detail through scientific studies sponsored by avid commercial interests. For most summers this century, there has been a well-staffed field research station in operation on one or other of the two main Pribilof Islands. It was there, in the summer of

1786, that the Russian Gerasim Probilof discovered the seals and the killing began.

The first Russian hunters on the Aleutian Islands called the inhabitants Aleutski, although they called themselves Unungan. Because the Russians were not as skilled as the islanders in hunting sea mammals, they captured and enslaved hundreds of them, transported them from the islands of the Aleutian Chain to the Pribilofs and set them to the killing of the seals. Within a short period sealing activities had twice nearly exterminated the herd, and in 1834 the Russian stewards took firm action. Under the first fur seal management programme they prohibited the killing of females, and the herd recovered. In 1867, however, the Americans bought Alaska and incidentally the Pribilof Islands; immediately huge numbers of females were slaughtered. Within ten years the sale of fur seal pelts, caught and prepared by Unungan labour, paid off the purchase price of the Alaskan territories, but the Unungans had no share in the profits and the fur seal herd was reduced to fewer than 300,000.

In one very tangible way the Unungans were now even worse off. At least the Russians had let them keep dogs, but one of the first acts of the American company, acting on behalf of the US Government, was the slaughtering of all pet dogs for fear that they might interfere with the seals. Today there are no pet dogs on the Pribilof Islands. Nor did Unungan health improve. The number of deaths had always outnumbered births and the population was still only maintained by the continual resettlement of more Unungans from the Aleutian Chain.

In 1911 the USA, Canada, Japan and Russia agreed upon a sealing convention which prohibited pelagic (open ocean) hunting and arranged for the sharing out of the land-based harvest of seals according to certain formulae. This agreement resulted in a rapid expansion of the Pribilof herd: it grew noticeably by about eight per cent a year, and by 1957 it was estimated that there were about one and a half million animals returning to the Pribilofs each summer.

In 1956 the managers of the fur seal herd decided that there were too many females and that this imbalance had led to low pregnancy rates. So it was decided to kill four-year-old females, and between 1956 and 1968 more than three hundred thousand were culled. Partly as a result of this action, the number of pups born each summer declined. More worrying, however, was the drastic decline in the size of the herd. The fall in numbers averaged between five and eight per cent each year, so that between 1956 and 1986 the total size of the herd more than halved. Fortunately this decline has slowed down in recent years (for instance, between 1981 and 1986 it measured only 1.6 per cent a year), and it may now have stopped.

Since 1984, no fur seals have been killed on any Pribilof island except to meet the subsistence needs of the residents, and on St George, where the story told in this book is set, commercial sealing stopped in 1972. Along with St Paul (the largest island) and three islets, St George makes up the group known as the Pribilofs. In the past scientists have used only St Paul seals as the statistical base for population analyses of the islands as a whole, while seals on St George have been the focus of attention for the study of the species' behavioural patterns.

The northern fur seal (*Callorhinus ursinus*) is the most ancient and distinct of all fur seals. Its generic name means 'beautiful nose', for it is its short muzzle which gives its characteristic profile. The maximum life span of a fur seal is thirty years, although on average females live to about twenty and males rather less. Most females produce their first pup between the ages of four and six, whereas males, although usually sexually mature by the age of five, are seldom able to acquire breeding territory until they are ten years old, and their active reproductive life is often only one or two years, rarely more than four. Females, however, continue to have pups well into their twentieth year.

The male northern fur seal weighs between 200 and 250 kilograms and is on average 210 centimetres long. The female is much smaller, weighing between 45 and 50 kilograms and measuring 140 centimetres long. The coat appears black or near black when wet. When dry, the coat of females and immature males is generally a silvery grey above and reddish brown below, with a paler chest. Adult males have heavy manes and a dark brown body colour.

As an eared seal, the northern fur seal is distinguished from true seals by its use of foreflippers as the principal means of propulsion through the water. It has abundant underfur and

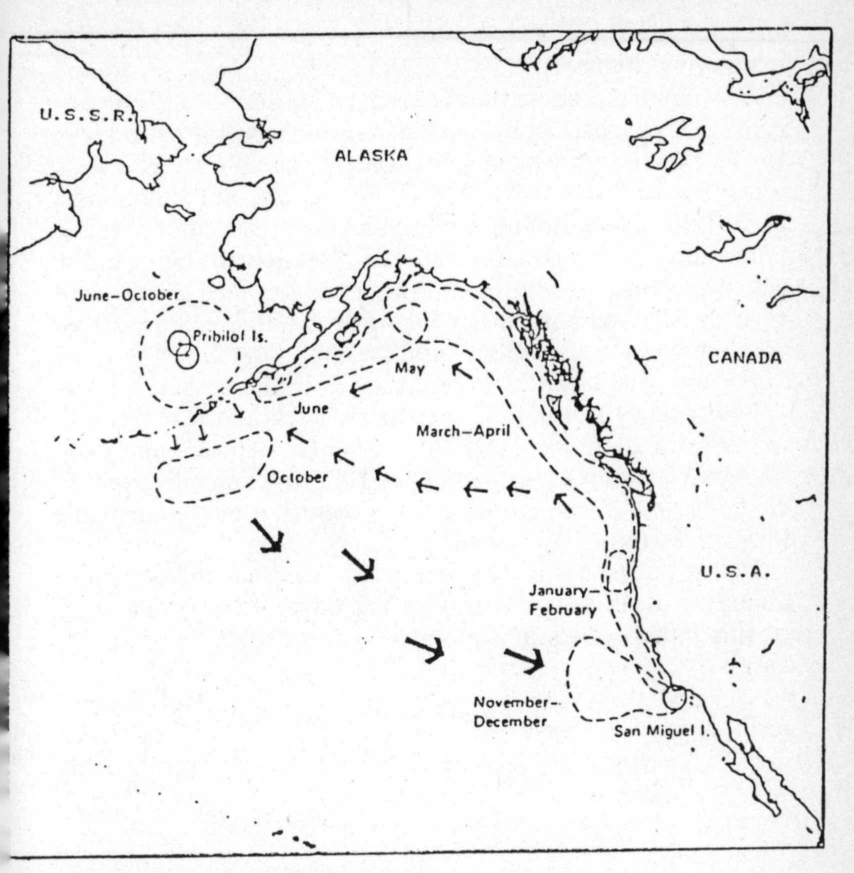


*Northern fur seal migration routes.*

tends to be associated with areas of cold water in the temperate oceans. The main population breeds on the Pribilof Islands in the East Bering Sea, but other breeding colonies occur on the Commander Islands in the West Bering Sea, on Robben Island in the Sea of Okhotsk, on the Kuril Islands in the east Pacific, and a very small, very recent colony has grown up on San Miguel Island off the coast of California. Except for the San Miguel colony, which probably remains in Californian waters all the year round, the seals leave their breeding islands in autumn and migrate south, the females and juveniles travelling farthest. They may be seen between 16 and 150 kilometres offshore, their usual range being between 50 and 110 kilometres. Unless they are sick or injured fur seals do not come to land, apart from visits to their rookery (the name given to the beaches on which they breed). Each summer the seals return to the Pribilof Islands of St George and St Paul to breed in crowded colonies. The conditions here are as close to being underwater as it is possible to be, without actually being in the sea: the islands are almost continuously shrouded in mist, and even in high summer it is cool, with Arctic breezes wafting damp air over dank beaches and swampy moorlands.

In June 1988 the National Marine Fisheries Service, the federal agency responsible for the research and management of fur seals, designated the fur seal herd as depleted under the US Marine Mammal Protection Act 1972. This new designation confirms public and government recognition of the fur seal's imperilled status.

I went out to the Bering Sea in that summer to learn more about the life cycle of the fur seal. What I found forms the basis of this remarkable story.

Paul Thomas
Radstock
1989

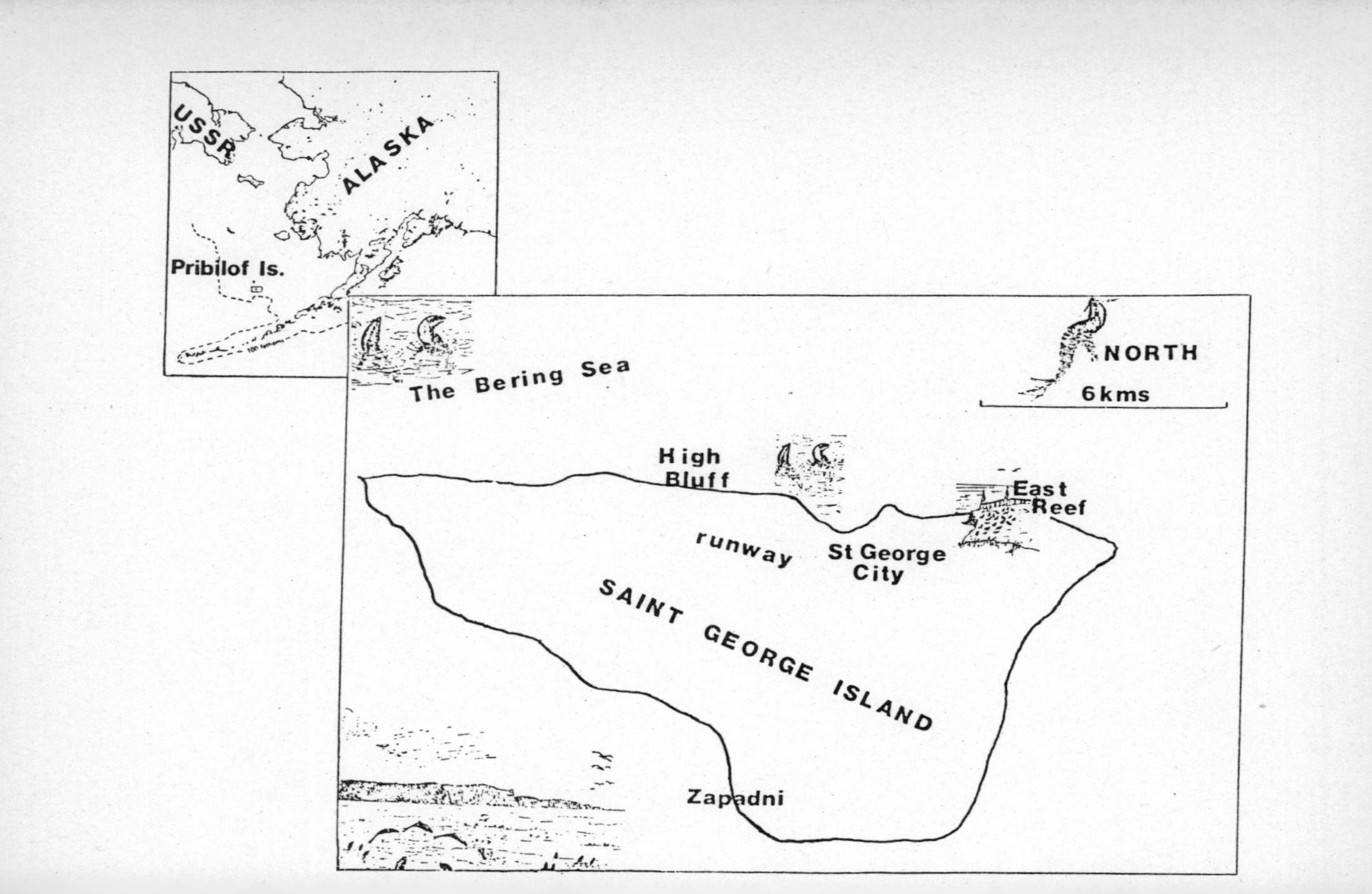
USSR
ALASKA
Pribilof Is.
The Bering Sea
NORTH
6 kms
High
Bluff
East
Reef
runway
St George
City
SAINT GEORGE ISLAND
Zapadni

# 1 EAST REEF ROOKERY

Darkness was complete. A tight softness held her deep in the warm waters which lapped noisily at her ears, in her mouth, in her nostrils—the almost metallic sound of a pounding, powerful pump and the water lapping. Irresistibly she was being drawn down, the constriction increasing. There was absolute power, totally controlling her small body. No matter how she wriggled and writhed she was forced down, and in her ears the almost deafening sound of her mother's heavy heart, beating faster and faster. In waves they came, those pushes and the pressure. She rolled, pulled her front flippers close against her chest, but the contractions were stretching her forward down a short tunnel, her neck, her head, elongated.

Suddenly, for a short moment there was a release, a rushing forward from her ears. The water about her was gone. The darkness was now a blinding red glow against her closed eyes. So cold! A shivering blast of near-freezing air assaulted her head, and around her throat was the choking tightness. Her mother moved forward and her head banged against a hard rock. She was half inside her mother and half on the rocks of the beach, being born. The noises of that cold northern shore were overwhelming her, the light boring deep into her closed brown eyes and the cold nipping keenly into the skin of her unfurred flippers.

Enormous male seals roared the roar of territorial possession, the clear calls of the females echoed against the bleats of a thousand seal pups; and, above it all, the crashing of the waves of the Bering Sea. This was her introduction to East Reef Rookery. And still the pressure, the pushing. About her face the translucent membrane, all that remained of her placenta. The vessels in her umbilical cord went into violent spasm. She was cut off from the life her mother had given her; her nostrils were empty, her lungs shrunken. There was no breathing, but with a final, shuddering slide she hit the grey boulder, stained black by the waters of her birth; fell limp onto her back, and slid into a small crevice between two smooth boulders. Great spasms in her arteries changed the direction in the supply of her blood. Her mother, above her against the grey sky, panted; called gently, clearly, shuffled carefully around, pushing her nose deep into the pup's soft side, and tbe coldness of the world entered the newborn's lungs. Her mother bit through the umbilical cord. Increasing blood pressure closed the connection between the left and right auricles. Within seconds she was a completely independent creature, no longer requiring the direct link with her mother. She turned, stood on her flippers, shook her head and was truly alive! Amik, named after the word chosen by the first human hunter to discover and describe the fur seal island.

Only yesterday, July 6th, her mother had arrived on that beach after a swim of five thousand kilometres, all the way from the Pacific Ocean off the coast of southern California where she had spent the winter months. In the waters of the blue Pacific, alone and solitary, the mature female seal had hunted and fed well; then in January she had turned north again, to swim to a group of cold, storm-rocked islands close to the Arctic Ocean. Now, on the island of St George, beneath a grey, grey sky, her pup was struggling to stand up straight, and calling—a plaintive bleat, high-pitched but rumbling: her own unique sound.

Suddenly there was a cold freedom; no more tight, dark constriction. She could move, do things for herself. Wet she might be, bedraggled perhaps, even wrinkled, but in her early strangeness there was a clear claim to the family face. She called, as all northern fur seal pups call. It was programmed into her mind. Cries between mother and pup are one of the very few social relationships fur seals demonstrate. Her mother was nuzzling her, pulling away the remnants of her foetal wrapping, calling gently. The pup rumbled back, sharp, high-pitched, whimpering and querulous.

She was *lacotha*, the word the islanders used to describe a fur seal pup, and as a pup she was black furred, whereas her mother, fresh arrived from the sea, was silver, sleek and shining. Other females, who had been on shore for four or five days, had dried out and were stained brown. They were close by her too, all crowded together, raucous and confused, heads twisting like snakes, bearing down into the crevice, their gaping mouths calling—not the call of her mother—and female breath, but not her mother's breath.

Then out of the sky her mother bent down with her own gaping mouth, teeth bared. She bit deep into the softness of the wet fur about the pup's shoulders and lifted the newborn, dragging her up out of the gap between the rocks. Another female leaned across, grabbed the pup's right front flipper and pulled at it. Her mother responded, pulling hard against the attack, then dropped her pup and snapped at the interfering female; roared defiant ownership. Amik bleated, for she was alone and cold against the hard unyielding surface. Other females roared; the scent of birth was strong, the air reeked with it. Seven females shuffled over, bore down on the pup, wove sinuous necks, their whiskers trembling in the cold breeze and their yellow teeth flashing bright in the gloomy light of the dull day. Amik saw them and cowered, trembling, blinked her big brown eyes. Again her mother called, but not a gentle, reassuring call: it was the roar of motherhood under attack. The other females retreated. Instinct restored them to their own programme as mothers of their own pups, cleared the confusion of their minds. They continued to quarrel for space, adding to the cacophony, but they had shuffled the short distance away from the pup to give her the space she so desperately needed. Now, with room to move, she searched hungrily, and her mother lay on her side, still and waiting.

The teat was black, soft, glistening amidst the drying fur. Amik licked eagerly at the dark, unfurred front flipper. There was no milk to be found there. Her tongue flicked out, worked its way up and across the female's body, found the teat; but as it did so, a neighbouring female roared, butting her mother in the neck, and her mother turned and roared back. The teat was lost to the pup's mouth. She gazed up at her mother who turned immediately, looked down at her pup and called gently, then edged back into position, and Amik found another teat ten centimetres above her first choice and began to suckle noisily.

Her mother's breasts were enormous: they ran the whole length of her body, a sleek, enveloping sac which did not interfere with her streamlined shape. Fur seal milk is the richest produced by any animal (at the time of birth it is over 50 per cent fat content). Tentatively at first, but with increasing confidence, the pup sucked hard and her mother's eyes closed shut while her back flipper waved gently over Amik's head.

Fur seals are nocturnal, most active at dawn and dusk. The pup was born in the dawning of that long Arctic day and she had her first meal as most of the seals on that cold shore quietened into long rest. Even the dominating males lay down amidst the wave-smoothed boulders, closed one eye and dozed. The pup heard the quietness fall, leaving only the bleating of pups and the lapping of the dampening sea. She changed teats and sucked hard again. The warm, rich fluid gurgled deep down inside her. Her nose was flat against the fur-covered blubber of her mother's belly and her nostrils were full of the scents of oil-filled fur and salt-rich air. It was the musk of her mother that was closest. She smelt it keenly, but there were other scents, too. The twelve females who claimed the right to share her mother's land all had their own unique scent, and the enormous male, who was four times the size of her mother, was redolent with his own odour.

Thus, although only minutes old, Amik was alive to her senses. Her ears picked up a light sound, padding across the rocks close to her, and her nose caught the stale whiff, a scent of foetid meat, pungent, very close, now still. She paused. Her mother, lying with her eyes closed, seemed not to notice, so the pup turned her head to investigate and saw it close, one and a half metres away, standing high on a rock looking down at her. Two sharp, keen brown eyes. Waiting. Watching. An Arctic fox in its summer pelt of untidy steel grey and brown fur.

The stare intensified. A thin tongue came out of a long mouth, sharp teeth grinned, the creature panted briefly. A male seal opened its second eye; yawned; closed one eye again and reverted to incomplete and ever-watchful rest. The pup's mother rubbed her head against her own neck, viewed the intruder from the corner of one eye and seemed unconcerned. The pup turned back, and as she found another teat the fox sprang, darted foward, causing a start amongst a few of the resting females who slithered nimbly onto their flippers and watched as the alien creature snaffled the afterbirth.

In less than thirty seconds the fox had licked the rock clean and was moving away, on sure paws, across the boulders of the shore, occasionally pausing on a big rock to look round, sniff the air and then continue. Although it was most adept and agile at jumping between the seals, the fox took care to skirt the great mound of a resting male before hurrying off into the distance. Half a mile away, in an empty overturned oil drum, five hungry fox cubs were waiting for more milk. Arctic foxes were common on St George, but unlike their mainland cousins, more than ninety per cent of the population was the blue rather than the white or polar form. Scavenging opportunists, they took almost anything, the seal rookery being just one place they visited. There, amidst the storm debris and the breeding seals, they searched diligently, looking for afterbirth or the carcasses of stillborn pups or diseased adults. Sometimes, however, they would climb the steep cliffs above the rookery and take eggs from the seabird colonies.

Suddenly the fox stopped, and stood on a broad yellow-painted line, about seven centimetres wide, which ran straight down one rock, up another and down again, where it was intersected by another bright yellow-painted line—in fact the whole of that part of the beach was divided up into a grid of squares, each numbered in ascending order from 1 to 100. The fox looked down at the line, then up towards the grassy hills which rose steeply from the beach; it glared, turned and ran at full speed, away to the west.

The pup, without identifying the noise, heard the unknown sound of a motorbike engine coming closer and closer. One male seal sat up, lumbered onto its front flippers and stood up to see over the slight rise which divided the beach from the grass of the mainland. This disturbed his neighbouring rival who also reared up and roared, and like a line of dominoes in reverse movement, the male fur seals of the beach rippled up into a rank of attention, roaring threats at each other and the intruder.

Abruptly the engine noise stopped. The scientist had returned. He parked his tribike about fifty yards from the beach, adjusted the hood of his parka, checked the buttons on his coat, pulled his gloves on more firmly and walked briskly to the ladder which led to his hide. To call it a hide was not really the right word: it was more like a garden shed on stilts, about four metres above the ground, with a very large window facing the beach and beyond it the sea. All the seals knew the shape and smell of the scientist. The hide, in fact, was an observation post to keep humans protected against the cruel elements. Even in summer the temperature on St George rarely rose above 11 degrees Centigrade and the clouds of mist hardly ever lifted completely, so to keep warm and dry the scientist had to build his hide. The door of the hut squeaked, opened noisily and the man disappeared inside, into silence. The male seals returned slowly to their resting positions, while the scientist got out his binoculars and notepad and checked the beach for new arrivals, marking the birth of the new pup against number 14, the grid square which he used to identify that particular part of the shore.

A brief, eerie calm invaded the seal rookery. All along the shore and just offshore, juvenile male seals groomed and preened, swam above and below the water in mock fights and attacks, practising their adult roles. Some of the eight- and nine-year-olds were waiting, watching vigilantly. If they suspected that one of the mature bulls was flagging or weakening in its defence of a territory, they would rush up the beach and fight, trying to usurp the breeding space and drive the old male into the sea. But it was not a territorial fight which attracted the attention of the scientist that morning, it was a sound which had all the seals looking out beyond the breakers—the long, low, bellowing of a right whale.

A small basking group of right whales had been washed in by surface currents, quite close to shore. Men called them by the unenviable name 'right' because the species was just right for hunting: the whale swam slowly, floated when killed and produced a high yield of baleen and oil. So it had been hunted rapaciously and its numbers had fallen to very low levels; even after years of protection from the whalers the numbers worldwide were still very low. Looking remarkably like floating crocodiles because of the concentrations of parasites on the tops of their heads, they had belched and moaned for more than an hour, blowing air through their nostrils at surface level or just above the surface of the sea. Three had turned, and in the shallow water close to the shore had begun leaping and slapping the surface with their tail flukes. Then, almost as suddenly as they had appeared, they dived, surfaced, dived and surfaced at increasingly farther distances from the shore.

An old male seal, close to a grassy bank farther up the beach, had watched the whales idly. He had returned to the beach for a third season—somewhat unusually, because most males held territory for one summer only. However, his land was not the most fiercely contested: it was twenty-five metres from the sea, occupying parts of grid squares 58 and 59, and only three females shared it. Even so, the season's battles and fasting had worn him thin. He had not eaten since coming ashore early in April and had lost over a quarter of his body weight. As he watched the whales he had, from time to time, even from that distance of twenty-five metres, caught the challenging eye of a very fit looking young male seal. The juvenile sat on a large rock amidst the breakers, and between bouts of grooming scanned the beach intently, staring purposefully into the eyes of any male seals, on territory, who deigned to return his glances.

The male on grid square 14, Amik's birthplace, gave short shrift to such young upstarts; he eyed the juvenile with such fierce hostility that the young seal knew it would be impossible to usurp that piece of land close to the sea. The winning of territory on the beaches of the Pribilof Islands was so important to bull fur seals that it determined their lifestyle. All winter, while sea ice locked the islands in its grip, the mature males swam in the near-freezing waters of the open Bering Sea, diving deep in the long dark nights of an Arctic winter to catch fish. There they awaited the spring thaw, while the females of the species swam five thousand kilometres south to fish the warmer, richer waters off southern California. Once winter retreated north to the Polar ice pack, the male seals hurried back to fight for territory on the beaches of St George and St Paul (the other main island of the group), and when a bull succeeded in gaining a patch of shoreland he remained on it, defending it and waiting for the return of females in early July, never leaving that patch of land, even to feed. So when the juvenile rushed out of the breakers and up the beach, the act provoked, on that part of the shore, a deafening cacophony of defensive aggression. Amik saw ten males bark furiously, jump to their flippers and give a firm and positive rebuff to the intruder. Small pebbles rattled noisily in the wake of the young male's rapid progress. Because fur seals have not developed their rear flippers for swimming they are very agile on land. Indeed, an adult seal can rush across a rocky beach more quickly than a fleeing man.

The juvenile moved directly and nimbly across those twenty-five metres of beach, evading contact with other males, until he was close to the grassy bank. The old male stood high on his flippers, across the boundary between grid squares 58 and 59, but there were no boulders from which he could repel the invader and now, in July, he was no match for this pounding young giant. The lithe, glistening body of the usurper darted straight at the old bull, who bared his teeth and roared defiantly. The younger male ignored this and bit deep into the front flipper, close to where it joined the body. That was the most vulnerable part of any male seal. The roar of the defending male changed to a strangulated shriek of pain as blood oozed out of the wound. Another wound close to the new cut had also opened up. The old male tried to retaliate, but the young seal darted out of the way and the old male slid unceremoniously down a large boulder into a crevice. The invader was at the back of his neck, but the thick mane-like fur protected the old seal who rolled over twice across the uneven rocks and, as nimbly as he could, stood up, turned and tried to steady himself against the full weight of a charge. The young male threw all of its two hundred and fifty kilograms hard at the throat of the old male. He resisted, caught the usurper a glancing blow behind the ear, turned and wobbled unsteadily down the beach, defeated and enduring further attacks as he ran the gauntlet of other males whose territories he had to cross before reaching the safety of the sea. Finally the old male seal slid into the breakers, leaving behind several blood-stained rocks and an eddying pool of pink water close to the boulder which the young male had mounted as he began his invasion.

The day progressed. The sun moved in its long low arc across the southern sky, and Amik dozed. Almost continuously, small trails of twenty or so seabirds flew back and forth about eighteen metres offshore, at a height of two metres, sometimes flying through the spray of waves breaking against the hard rocks. And on a promontory a group of twenty cormorants preened and groomed, while single birds flew off from time to time, to skim the surface of the cold ocean, alight and bob and dab in the unbreaking waves, before submerging and flying through the water in search of prey. From the distant hills came the faint sound of a raven's cry and a snow bunting chick cheeped from the wooden framework supporting the scientist's hide. Inside the small hut, the man opened his lunch box and began eating his sandwiches. Around him were the tools of his research: a notepad with columns neatly drawn for recording the presence and absence of each marked seal, and another notebook for recording the happenings on the yellow-painted grid squares; there were index cards, recording previous seasons' findings, binoculars and a small field telescope. Dr Alan Steinberg was one in a long, long line of scientists who had sat on this beach recording. He had first come to St George seventeen years ago and had returned every season since, to rebuild the hide on the foundations laid at the beginning of the century by previous research teams and to observe and record. He had seen the size of the fur seal herd decline. Now he was trying to discover the reasons for the decline, whether it had levelled off and, if it had, why.

Out at sea the noise of a group of five glaucous gulls attracted Dr Steinberg's attention. They were big birds, as big as a great black backed gull, sixty-eight centimetres long, and it was the greyish-blue plumage on their backs which gave them their name. He put his sandwich down and picked up his binoculars. The birds' presence had dominated the pup's day. They had flown around the right whales when they were close to shore, feeding on the shrimps the whales forced to the surface, but when the whales had swum off the birds had returned to the shore. There they moved noisily between the sleeping seals. Although not as vociferous as some gulls, they uttered a variety of wailing notes similar to the call of the herring gull. When they visited the rookery they often searched out faeces, picking up nutrients still contained in the seals' waste products. Glaucous gulls were great opportunists, always looking out for additional food, so they were easily distracted by the threshing of a stellar sea lion offshore. This creature, a direct relation of the fur seal, had picked up a piece of halibut thrown overboard by a passing fishing boat.

Dr Steinberg reflected that there were fewer sea lions about than there used to be. He had watched as their numbers had declined parallel to those of the fur seals and other creatures dependent upon the waters of the Bering Sea around the Pribilofs. The gulls flew out to harry the creature as it swung its head from side to side, trying to break off manageable pieces of fish flesh to swallow. The scientist focused on three gulls in particular. They flapped directly over the sea lion's head in an attempt to get a piece of fish for themselves. The sea lion was doing its best to feed and fend off the birds at the same time. It was not easy, but a Peninsular Airlines plane flying in from Anchorage indirectly helped the struggling mammal. As the plane swooped in low, to land against the westerly wind, the gulls took flight and flew off, leaving the sea lion to dive, still holding the halibut remnant firmly in its jaws.

Dr Steinberg scanned the beach to see if the noise of the aeroplane had disturbed the rookery in any noticeable way. No, it had not. He noted the time, the type of aircraft and the fact that the seals were not disturbed. This was important information. Since the seal harvest had been stopped in 1972, the islanders of St George had been left without a livelihood. The US Government had given them a grant of twenty million dollars and with this money they were building a big new harbour on the south side of the island. Once it was finished they hoped to attract Japanese and Taiwanese factory fishing boats. Part of this development included a new runway large enough to take medium-sized jet cargo planes. These would be needed to import supplies and bring relief crews in and ferry out fishermen going off on leave, but before such a plan could be put into operation permission was needed from the major conservation bodies. Dr Steinberg worked for one of them. He favoured extending the existing runway close to the settlement of St George, rather than a completely new airport on the other side of the island. However, it was not personal preferences the Government agency wanted; what was needed were facts about the impact on marine and terrestrial wildlife from human activity —hence the scientist's interest in the arrival of the plane.

Suddenly Dr Steinberg murmured a mild curse and began

packing away his equipment. The plane, he had remembered, was bringing his research assistant for the summer, an undergraduate from Seattle University, his home town. She was going to help him collect data. He hurried out of the hide, mounted his tribike and rode off in the direction of the airstrip.

Six and a half kilometres away, to the west of the rookery, there was a tumbled rock-strewn cliff face—coarse grass intermingled with sheer outcrops and scattered boulders of all sizes and shapes. The northern slope of Ulaka Hill swept down to a wide plateau of moor and marshland at the centre of which lay Seal Lake—not that any seal had ever climbed up to bathe in its fresh waters. This plateau lay just under seventy metres above the level of the rookery, but it was Ulaka's long eminence which brooded over the airstrip. The hill itself was about two kilometres long, running east to west and rising to nearly three hundred metres in height. For thousands of years least auklets (the smallest of the auks, only sixteen centimetres long, related directly to puffins and guillemots) had returned to Ulaka's slopes to nest and roost in their hundreds of thousands. White stains lay in clear contrast to the lichen and moss which clung to the dark volcanic rocks. Now, however, this characteristic white staining was limited to a three hundred metre stretch of hillside. To the east lay many rocks with the white stains of previous occupancy fast fading, and many more rocks that had once supported roosting birds, at densities of over a hundred to a metre, were pristine black and lichen-covered, with no evidence of least auklets at all. The bird colony was now one tenth the size it had been thirty years ago.

The plane landed, relatively smoothly, on the island's central plateau. The wide ash-covered runway also served as the island's main road and cut a brown swathe across the dark green marshland of the interior. The airport terminus was a small wooden shack, and as the plane taxied back several islanders, including Alex Merculieff and his father, Vladimar, hurried forward and helped unload small items of equipment, mail and fresh vegetables, all flown in from Anchorage on mainland Alaska. Alex often worked with Dr Steinberg and knew that the research assistant was due on the plane that day. In the absence of the doctor he welcomed Elizabeth.

Within half an hour the plane had turned and flown off, but

this time its flight path westwards was not over the rookery; instead it flew close to the cliffs of High Bluff, a vertical rock face which rose two hundred metres straight out of the sea. There, thousands of common murres were brooding their eggs. In Britain these black-backed, white-bellied birds, with elegant necks and long pointed black beaks, are called guillemots. They breed in bodily contact with their neighbours, at greater densities than any other bird species. On those high cold cliffs overlooking the Bering Sea, they were packed thirty pairs to the square metre, although in many places the density was not what it had been thirty years before: then there had been ledges and crannies where the density sometimes reached forty pairs to the square metre. Each female laid a single egg onto the bare rock. Such high-density grouping presented an impenetrable barrier to predators such as the glaucous gulls, and the murres were safe from the island's foxes because even those sure-footed creatures could not climb the sheer rock face. Thus, most years, murre pairs reared their single chick successfully. The noise of the aircraft engine unsettled them, however, and many flew anxiously away from the cliff face, veering out in their thousands over the dark sea before turning sharply in the wind, and only as the engine sounds faded into the distance did they return to their eggs.

Dr Steinberg had reached the runway by the time all this disturbance was taking place. He knew that aeroplanes disturbed the birds, and this fact would also be taken into account in the final decision regarding the siting of the new or extended runway; the welfare of bird life was not his direct responsibility, however, for his brief was to research and report on marine mammals, and in the case of St George that meant fur seals and stellar sea lions.

It had been Dr Steinberg's intention to ask Elizabeth to stay in the skin plant, the large brick and weatherboard factory building where seal pelts had once been treated in their thousands, and prepared for the world market. For six or seven years after the harvesting had ceased it had stood idle, used only as a glorified workshop and storage shed. Then Dr Steinberg, during the heyday of his research in the early 1980s, had obtained funds to build a self-contained four-bedroomed flat in part of the upper storey, for use by his assistants. Recently, however, the decline in funding had resulted in few people staying at the skin plant. The research station itself was a large weatherboarded cottage, similar to those in which the Aleut islanders lived. It was warm, comfortable and home to Dr Steinberg for nearly half the year. His other home was close to his laboratory in Seattle, where the government department for which he worked was based. Money from the US Fish and Wildlife Service had helped fund some of the work the islanders did, and this was especially true of Alex Merculieff, who worked not only as Alan's assistant, but also as tourist guide and guardian of the fur seal herd. He was also in charge of the island's one and only bus, a twelve-seater, four-wheel-drive vehicle which was used in the main to ferry goods and equipment from the airport to the village, for airline passengers were few and far between and all the islanders had their own cars or trucks. As the four of them drove back to St George, it was decided that, as Elizabeth was the only assistant that season, it would be easier if she stayed in the spare bedroom in the field research station.

On the rookery, Amik's afternoon doze had been disturbed by the new territorial male who had begun to occupy squares 58 and 59. He was evidently nervous of his new-found power and had barked noisily and patrolled his territory ostentatiously for most of the afternoon. Although neighbouring males found this irritating, none was strong enough to silence him. The pup watched intently as the short, ritualistic challenges were repeated all around the new bull seal's territory. They were energetic flurries of possession, but because the youngster had proved his right to hold territory, none of the interchanges approached the ferocity of his earlier battle for that small portion of the shore. The new male also had to establish his right to share land with the three females. They had been unsettled, and barked and snapped at him as he approached; none of them was in heat. Instinctively he knew that, because their breath smelt the same as usual. When a female fur seal comes into heat, the smell of her breath changes and for a maximum of thirty-six hours she is receptive to the advances of a male. The defeated old male had served these females and that, in part, explained why he had capitulated to the challenge of the new young male.

As evening approached and the new male began to settle down into longer periods of quiet inactivity, a keen ear would have heard the sound of slight scratchings coming from a crack between some tumbled boulders, not far from that same grassy bank where, earlier in the day, the two male fur seals had done battle. Suddenly, without warning, a four-inch collared lemming came scurrying out of that gap between the rocks. It was a female, emerging from a grass and lichen-lined nest where she had left her litter of six, her second brood of the season. She paused on the boulder and sniffed the air—not only for danger. The solitary animal had marked her territory carefully with a special scent, and she now searched the breeze for clues which might indicate another lemming trying to take over her land. Not even males would she tolerate, except her own mate, and even he could not come close when the litter was small and defenceless. Her small brown eyes twinkled sharply, her whiskers trembled; she lifted her throat, stretched her nose into the air, took three short sniffs, and plunged headfirst down onto a small piece of overgrown beach. Her front right paw carried stalks of coarse grass to her mouth while she sniffed the ground with her nostrils close to the earth. Earlier in the spring she had shed the extra-big claws on the third and fourth digits of her foreleg. She had grown them in the autumn for digging in the winter snows, and shed them each year, in the spring, when she no longer needed them. Nervously attentive all the time, her tiny ears, hardly noticeable above the thick brown coat, listened for sounds of danger. Immediately above her head, yellow Arctic poppies trembled in a cold breeze and a single chocolate lily attracted a few insects, but the only real drone was the sound of sleeping seals.

Amik lifted her head to get a better view of the lemming which was no more than a blurred brown illusion on a distant boulder. That slight movement was more than the lemming could tolerate, and she darted into a small clump of lavender lupin, just as the new young male jumped into action, scurrying off the beach and onto the grass. He chased Elizabeth as she came out of the hide and climbed down the steps. Alan had brought her out to the rookery on her introductory visit and had left her alone in the hide for several moments as he checked other parts of the beach. She silently cursed herself for not being more wary. She felt quite embarrassed, almost stupid, for being so frightened, but Dr Steinberg, who was waiting for her at the jeep, simply laughed, then fell suddenly serious and explained to her that bull seals were dangerous when they were on territory and new male seals were worst of all and had to be watched very carefully. So the young undergraduate took note of what her supervisor said, and then both of them returned to the research station which was their temporary home base. However, the roar of the male seal was the trigger for the awakening of the beach. With the starting up of the jeep's engine came the call of a hundred deep-throated territorial mammals at the height of the breeding season, all declaring their ownership rights.

As the vehicle stuttered into top gear across the low grassy plain, the pup got her first close sight of the male of grid square 14. He came on his regular inspection of the females grouped on his land, smelling their breath, searching for females on heat. He got short shrift from Amik's mother: as he descended from a boulder and came close to her face, she snapped, growled and howled at him. He retreated.

The beach was suddenly bathed in a warm, weak light: for the first time that day the clouds had parted and let through a low twilight sun. It was the first time Amik had seen the sun. It was warm. The pup blinked, half-closed her eyes, tried to identify the source of the light, but it was too harsh, and, within moments, the clouds had scudded back into place. The sun was hidden again and the usual Probilof weather returned, with its heavily overcast sky. The fox returned too, padding along at the top at the beach. She sniffed low and long, close to the lavender lupin and at the crack between the boulders. The lemming had long since disappeared back into her nest, but she smelt the smell of fox and, trembling, quietened her brood until the shadow had passed. Amik waited until the fox continued her jog towards the east, before she returned to her mother's teat and quietly suckled.

# 2 MIST AND MURDER

The long white fringe of spray, waves breaking against the island, went unseen as the sun rose. Mist blanketed most of St George and Amik lay wet on the hard rocks, asleep. It was not the darkness of the womb she had experienced during her first hours, but it had been a black night. The fog had fallen quickly and heavily, not coming in from the cold Bering Sea but drifting down from High Bluff, ten kilometres to the west. Mist had covered the cliff from late afternoon of the previous day. It had crept down in clouds not long after the research scientists had returned to base for supper, following Elizabeth's fright with the new bull.

The tiny seal had not noticed the mist because there was no clear sight line to High Bluff from East Reef Rookery, especially for Amik who was just over sixty centimetres long and lay close to the jumbled boulders of the beach. However, she had heard the mist approach. As the sun dipped below the horizon and the clouds rolled heavy and white down to the sea, the clamour of the sea birds had quietened and the echo of waves across the rocks, the sharp clatter of pebbles knocked aside by females scurrying into the water, became muffled and deadened.

In the gloom the whole of the rookery and its activity seemed muted. Heavings, grunts, groans and only the occasional roar, broke into the monotonous sound of the pup's tongue flicking and licking, pulling and sucking at her mother's teats. She did not see the scores of females leaving their pups to go out to sea: undisturbed, apart from the discomfort of cold hard rocks, she had suckled and rested, suckled and rested, and felt the warmth of her mother's belly. Over 60 per cent of the female's total energy output was directed towards her, and the pup guzzled on and off all night. As she ingested the milk she also swallowed several tiny larvae, just over half a millimetre long.

Her mother's milk was contaminated and from birth Amik was being infested through the very liquid which gave her life and sustenance. Hookworm larvae had remained dormant on the shore of St George since the previous year, when other infested pups had passed faeces containing eggs of the parasite. The eggs had hatched on shore, moulted through two stages, and in the third stage their development had been arrested over winter. Somehow the tiny creatures, almost invisible to the naked eye, had survived, and with the coming of summer and the body warmth of the adults, had begun to move across the rocks with wavering heads, upwards, always upwards. They found the naked skin on the flippers of the fur seals, penetrated the tissue and, travelling through the blood or lymph glands, reached the female's breasts. Once in the milk, the pup naturally ingorged the hookworm larvae and within hours of her birth these insidious parasites were in her gut, moulting again, this time into adulthood. Then as tiny worms, only one and a half millimetres long, they began to feed on the blood of the small seal by attaching themselves to the wall of the intestine. Once mature they mated, produced eggs and thus completed the cycle. Acute hookworm infestation could result in death from anaemia, but by some quirk of evolution, the adult hookworms were found only in fur seal pups; no mature fur seal suffered any ill effects from hookworm, although many adult seals, both male and female, carried third stage larvae in their belly blubber.

The fog remained well into the morning, and it was not until the sun was quite high that its weak rays finally brought some light to the beach, making the air white and swirling. The cold wetness continued to dampen sounds, but even the thickest fog was unable completely to blot out the cacophony of sea birds calling. St George had one of the largest sea bird colonies in the northern hemisphere. Each year millions of birds returned to its misty shores, and each summer morning hundreds of thousands left their roosts and nests on the high cliffs and flew out in long lines to hunt the open sea. Nestling deep in the warmth of her mother and shrouded in that thick mist, the pup heard them going, the loud whisperings of the wings and the raspings of the gulls' cries.

Occasionally she paused in her feeding, climbed up her mother's side and looked around. All she could see were jumbled black boulders and the brown shapes of other female fur seals resting close by. Her mother was still a silver seal, but already the shining steely clarity of her coat was fading. The pups, with their jet-black fur, were difficult to identify amidst the hundreds of wave-smoothed boulders on the shore. Amik slid back down, found a teat and began to suckle again, eyes closed, tight to her mother's fur.

Out of the white air came the lumbering sounds and heavy pantings of the male. He was on patrol, checking the females, sniffing at their breath. The pup turned her head and saw him coming, while over the side of her mother's sleeping form she saw another male looming up out of the gloom. Her mother was a young female, only five years old, and she, Amik, was her firstborn. So her mother's tiny territory was at the edge of the female group, close to the boundary of the male's territory, close to the land of another male. Thus as each male checked the females on his land, they came close to each other.

Those males had sat on that beach, as uneasy neighbours, for the last three months. They knew each other well. However, defence of the territory was not entirely a performance ritual. It was real, an instinctive evolutionary programme from which no seal could deviate. They came against each other in that swirling white fog, over the bodies of the young female and her firstborn. The neighbouring male roared. The pup flinched and her mother sat up. Much of the defensive behaviour was mere posturing, to conserve energy, but if in its performance there appeared to be a chink in the determination of the other, one aggressive male would seize the opportunity to challenge the right of his neighbour to the females on his land. Injuring a female in the mêlée was of no concern to the males; the overriding instinct was defence of a territory.

Neither of the males had sired the pups being born on the shore that year. They were the new owners of the land close to the breakers but not flooded by the tide—the best land—and they were determined to defend it unflaggingly. As they came close they roared, barked sharply, shook their heads fiercely and violently and stared obliquely at each other, with an awkward turn of the neck. The female backed off, only to find

e bleak northern coastline of St George.

'City' of St George clustering below the towering cliffs of High Bluff.

Every summer the female fur seals gather in their own territories on the cold boulders of th
shore, and the males fight to hold the land.

A bull seal roars his challenge to would-be usurpers.

mature bull seals gather on the former hauling grounds, where seals were once slaughtered r their pelts.

e older juveniles wait at the sea edge for an opportunity to seize a territory of their own…

And every female leaving or entering the water must run the gauntlet of their ardent importunities.

Some bulls prefer to seek territories higher up the shore, where there are fewer females but a fewer challenges to their ownership.

herself being jostled and attacked by a neighbouring female who felt that her private space was being invaded. The pup saw her mother move away and felt a sudden sharp pain in her back as a loose pebble hit her. The shifty-eyed but drawn-out exchange of glances between the males had erupted into an unexpected lunge, sending a shower of loose rocks and pebbles across the beach. Amik scrambled to her mother's side, but the young adult female was still exchanging head-twisting and barks with her neighbour and momentarily ignored her pup, while the males had reverted to more violent head-shaking and barking.

It all seemed very close and very threatening to the small pup, gazing up from her lowly position. There was a frightening intimacy implicit in the challenges and struggles in which all the adult seals seemed to be involved, and shrouding clouds of fog intensified and added menace to the aggressive, overwhelming shapes of the adults, especially the males. The bulls lunged at each other again across their invisible dividing line, endeavouring to keep their vulnerable foreflippers as far away from each other's mouths as possible. The thick skin and manes on their chests and necks fended off any potentially serious blows. Another flurry of pebbles rattled against the larger boulders and then, as suddenly as they had begun, in a gruff, almost reluctant way, they turned their backs on each other and continued with the stock-check of females on their premises. Immediately the pup's mother moved back to her spot against the big black boulder, and Amik followed.

Peace was short-lived, for within seconds the male had returned to the female and was breathing hot and lustful down onto her, but she snapped back, repelled him again, and again he retreated to his favourite resting spot in the centre of his land, where he could lie with one eye open and watch. Thus, as the fog slowly lifted and calm was restored to that small corner of the rookery, the seals settled down into a fitful rest.

The tide was out, but because there was little tidal variation on St George, there was no obvious increase in the amount of beach available for males to colonise. A line of sea urchins, about one and a half metres wide, indicated the extent of additional land, and crunching over these, a late-arriving pregnant female struggled up onto the rookery. The closest male stood up, barked, and moved forward to greet her in the usual aggressive way. She did not snap back. A deep infected would in her side had drained all energy from her. Somewhere out in the deep ocean, several months ago, she had been injured—perhaps it had been fishing tackle or perhaps she had swum too close to a propeller, or escaped with her life from a pack of marauding killer whales. Whatever the cause, not all the salt in the sea had healed the wound. So she pulled herself over the rocks very slowly.

The male pounced on her, bit into her back, lifted her up in his jaws and swung her through the air, carrying her towards his territory; but an angry bellow behind him was followed by a great tug at the female. Another male had caught hold of her tail and was pulling in the opposite direction. The female seal shrieked in pain, writhed desperately, bit into the cold air. To no effect. The first seal, suddenly, sharply, tugged again and a jawful of seal blubber filled his mouth. The opposing seal recoiled uncontrollably with the female seal's tail in his mouth and the rest of her body banging mercilessly across the hard rocks. She twisted quickly, tried to bite her attacker, but could not reach him. The other male spat out the blood and blubber and lunged forward to grab at the female's neck. He pulled again. This time the female was close enough to bite back. She sank her canines deep into his neck and blood oozed over her face. For a moment the bull let go and she was a swinging hammock, her tail held in the jaws of one male and her teeth buried in the neck of the second.

All the fighting, confusion and noise had attracted two other males and a juvenile out of the waves. Each of them dived, lunged, bit out to secure a piece of the female. This was too much for the first bull on whose land all these males were fighting. While they tore at the female, he laid into the juvenile who retreated immediately into the breakers and swam off shore, to turn and watch the remaining males withdraw, one bearing the carcass of the unfortunate female. The first male tried to enter the territory of his neighbour to rescue the corpse which had been dumped unceremoniously amongst the other females, but failed. As the beach quietened and the roaring, barking males lay down again, the victor moved across to smell the breath of his newly acquired female. But there was no breath to smell.

It was another example of the ferocity of the male seals. Dr Steinberg had told her about it in the jeep on the way home as they talked about the safety rules for the rookery, but Elizabeth had in some ways dismissed it as part of his exaggerated anecdotes about fur seal life. Now, as she noted down all the details—the number of the grid square, looked through her telescope to identify the marked male seals involved—she could understand why her supervisor had said what he had. The telescope was always used for very close work, like identifying tag numbers; binoculars were reserved for overall general observations. And while Elizabeth focused up close on the males she thought back to the fright of the evening before. These seals, at least the bulls, were dangerous and she must be more attentive and careful when approaching the beach.

The dead female lay on her side. Elizabeth could see the animal quite clearly through the telescope, but the cow lay on the flipper where the green tag with its important number was fixed. Normally the observer would wait, keeping an eye on the seal until it moved into a position where it could be identified from the number on its tag. This seal, however, was dead; it was not going to move. Elizabeth could of course tell Dr Steinberg later in the day, and ask him to come and help her move it. But what if the fox came and took the tag. . . ? No . . . She could radio and ask what she should do . . . No . . . She was a scientist, she would go onto the beach and check the flipper tag herself, now, while the observations were fresh in her mind and incomplete on her notepad. The pup heard the wooden door creak on its hinges, saw the scientist climb quietly down the ladder onto the grass. Elizabeth walked warily across the few metres of unoccupied beach which lay around the base of the hide.

Inside the small hut the radio crackled into life. It was her supervisor.

'Liz, just to tell you that Alex is on his way down to the rookery.'

There was no answer. The radio lay on its side on the narrow wooden ledge, unattended.

'Liz. Liz?'

Through the open window of the hide, Elizabeth should have heard his voice had she been in a fit state for comprehending such things. She was not. She was trapped between territories, speaking quietly to two males. Each bull eyed her and eyed each other. She was saying:

'Now, you guys, don't worry. I'm not a seal. I'm just a little old research assistant, just come to do a job.'

She was thinking of all the things that she had heard in the jeep the night before: 'Don't turn your back on the critters, they can run faster then you. Just outstare them. Nine times out of ten they'll give way.'

And the tenth time? And the tenth time? She could not remember.

Amik heard the soft-pitched whisper, saw the three shapes from her low position on the rock. They stood against the grey skyline: two brown-maned bulls waving their heads and barking, and a thin, erect figure in dark green, just a circle of pink flesh peeping out from between the folds of her hooded parka.

'Now quietly, you guys,' the scientist was saying.

'Hey!' The shout frightened her. 'Get out of it!'

Two men. Smiling broadly! The bulls turned and lunged up the beach towards them, but they were well back from the seals' territorial limit, standing high up on the grass. The bulls rushed to the edge of their defined land and tottered with threat lunges and roars.

Elizabeth felt more confident.

'You guys frightened the life out of me,' she shouted up. 'I'm just getting a tag number.'

She rushed down to the dead female, lifted the flipper, noted the number and hurried back, chased in a half-hearted way by another bull, whose territory she had not quite skirted. Breathless, she jumped up onto the grass, pulled down the hood of her parka and smiled broadly. 'How did you get here?'

They pointed to the two-seater four-wheel motorbike parked on the grass close to her tribike.

'The quadribike,' smiled Alex. 'Liz—' suddenly he was very serious. 'I know you're new and don't know everything yet, but you're not supposed to go on that beach without somebody else with you.'

'I was just getting a number,' she explained. 'The corpse. The bulls killed it. Tore it apart. I needed the number.'

'Like you need a hole in the head,' chided Alex. 'They're savage, these guys. You should go careful. Where's your club?'

Leaning up against the wall in the corner of the hide there was a stout piece of pine, almost like a baseball bat.

'Don't believe in that,' she said. 'Not even for protection.'

'Well, don't get yourself into a corner.'

'I wasn't. Dr Steinberg's told me what to do. If they challenge just outstare 'em. Never fails to work.'

'Dr Alan's used to them, got confidence. It's not so easy.'

Alex's father had said nothing. He was overawed by his son's familiarity with the scientist.

'Mind if we come up to the blind for a while? Give 'em the once-over?' Alex used the American word for 'hide'. He was not a researcher, he was a protector, paid a retainer to check that the seals were not unduly disturbed.

Elizabeth nodded and smiled. Alex winked at his father, slapped Elizabeth across her back and she led the way up the steps. Her heart was still pounding furiously. It was then that she heard Dr Steinberg's voice on the radio, calling her.

The pup suckled all morning on and off. Already the fat content of the milk was declining as her mother's energy stores were sapped, but the pup was strong and growing almost visibly. Another pup was born close by and, in a flurry of agitation, a gaggle of females quarrelled and investigated over and around Amik, but by clever dodging and weaving she managed to avoid being trodden on.

The two men sat, almost silent, in the small hut for nearly an hour. It was not an embarrassing or threatening silence. They were watching, with intelligent interest. Occasionally Alex would describe an aspect of behaviour, for Elizabeth's benefit. She was impressed with their patience and their observation skills, but she was pleased to be able to show off some of her own knowledge, too, when a fair amount of disturbance to their left marked the progress of a bull hurrying down to the sea's edge. It appeared to be drinking.

'Well I'm jiggered,' whispered Alex. 'That critter's drinking seawater.'

They both looked at Elizabeth.

'Gee, that is unusual this time of year,' she exclaimed. 'Bulls drink seawater when they're switching from protein to fat-based metabolism. You know, getting ready for fasting. These males sit on shore guarding their territory for anything up to sixty days. That's a new one, only took up position yesterday, so I suppose that's what he's doing.'

Alex seemed impressed. He nodded with appreciation and the group fell into silence again and watched the scene through the glass of the hide's window. There were more stellar sea lions off shore and harbour seals, engaged in territorial displays just out from the fur seal beach, and all the while hundreds of murres were streaming by, flying in long, low lines out to their fishing grounds.

Elizabeth pointed out another birth on grid square 14 and noted the appropriate facts in her own records. Not long afterwards the men decide to leave, but not before warning her not to go onto the beach again unless there was someone else with her. She walked out of the hide with them and climbed down the ladder. She stood beside the red-painted quadribike and thanked them for their advice, asking them not to mention it to Dr Steinberg. Inwardly she felt furious for having been caught in such a vulnerable position. As she waved goodbye to the two islanders she heard the jeep coming over the hill and down towards the rookery.

The grey jeep bumped and juddered down the rough track. Juvenile seals stood up, huddled together, watching quizzically as the brakes squealed and Dr Steinberg brought the vehicle to a halt, quite close to the catwalk. He jumped out and walked across to her.

'How are you getting on?' he asked.

'Fine. Fine. Just fine.'

'I've brought a better staircase,' he said. 'Made it this morning. That old ladder's too rickety.'

They set about unloading the staircase. The snow bunting chicks were still chirping noisily in the grass and fluttering up onto the debris of broken hides, destroyed by the violent winter storms of previous years. They had been hatched in a deep crack between three boulders at the top of the beach, not far from the hide, in a comfortable nest lined thickly with fur gathered from the seal rookery. Both birds were in attendance. The male, distinctive in his summer plumage of black back, white head and belly, perched on a stout dead stalk watching the human activity, while his mate, in dowdy brown and dull white, was frantic lest the humans take her chicks. She made a show of noise and display, flying down onto the path in front of the jeep with much wing flapping almost in front of Elizabeth's feet, trying to lure the humans, distract them from finding her chicks. Despite her histrionics, neither of the scientists had seen her chicks: they had eyes only for their task and for the seals, and were busy carrying the staircase to the edge of the beach.

There, having pulled away the ladder and set everything up, Dr Steinberg stood, bare headed, with the wind blowing through his hair, and began hammering nails into the wood, fastening the staircase securely to the hide. Elizabeth stood watching the seals, her eyes picking out the gaggles of females dotted all over the rocky shore. The scientist seemed to have noticed what she was looking at and commented quietly:

'They're not harems. Females share the shore with males because it's a good place to breed. The best places to breed are those most attractive to females, and therefore the males when they're juvenile notice where the biggest bunches of females congregate, and fight hardest to win those places. So the strongest males occupy the best land and that makes for the best pups and the highest survival rates.' He paused, and hammered home a few more nails.

'What about those?' Elizabeth pointed across the grass to East Cliff Rookery, two hundred and seventy-five metres to the east. There on the narrow rocky coastal plain, covered in storm-tossed abandoned tree trunks, another seal rookery teemed with life, but high up the steep, turf-covered cliff face were several males, well spaced, almost isolated.

'Yes,' he said. 'Not all males choose to fight it out close to the water's edge. Some climb high above it all and lord it from up there.'

'Why's that?'

'Well, they don't attract so many females, but then they won't have to expend so much energy in fighting, so maybe they'll come back for several seasons while the bulls on the main shore are lucky to last more than a single year.'

High on that distant grassy bank a sheer black silhouette stood motionless, looking down on them. Elizabeth picked him out: one bull seal which had chosen to make his territory well away from the beach.

'Probably born somewhere up there,' mumbled Alan, his attention reverting to the staircase. Elizabeth, however, knew what he meant. He had already told her that most seals returned to sites close to their own place of birth.

'This catwalk,' he said, 'is getting old, too. Although it's not really needed any more.' Elizabeth nodded. She had not bothered to climb up the ladder close to the roadway and walk along its planks. 'Time was,' continued Alan, 'when I first came here, there were seals all over this grass.'

'Really?' Elizabeth was impressed. The catwalk was a good two-and-a-half metres off the ground. It began at the end of the roadway where she parked her tribike and Alan left the jeep, and ran thirty or forty metres across the rough grass to the hide which stood high above the rocky beach where most of the seals crowded at the edge of the sea. Made from rough but neatly prepared timbers—all of which had been imported because St George grew no timber at all—it was an impressive edifice on that wild, desolate shore. From a distance, driving down the road from the central plateau of the island, it seemed very spidery, fragile and untidy; now, from below, she could see that with its solid timbers, strong enough to take the weight of two or three men, it was a bridge designed to allow scientists to creep forty metres back to the hide without disturbing any seals.

'Yep,' mused Alan. 'Seals were close-packed all over here—' it was as if he knew what she was thinking. He pointed to the wide grassy area which fringed the entire length of the rookery. 'The only way from the road over there was over their heads. That's why we built the catwalk, to get to the blind without upsetting the seals.'

Elizabeth considered. 'So the decline's really noticeable?'

'Oh yes. There. That's fine.' He stood back and admired his handiwork. 'Much easier for you, getting in and out of the blind.'

'Do you need me now?' asked Elizabeth.

'No, that's OK,' replied Steinberg.

'I'll get back.'

'Be the first to use it,' he smiled, looking at the staircase.

'Yes,' she chuckled. 'Oh, by the way, over there. There's a dead female. The males killed her when she got on shore.'

All the activity at the edge of the rookery had unsettled the males. Although it was mid-morning and therefore a rest period for seals, many of the bulls stood up on their flippers and roared as Dr Steinberg moved across the edge of their breeding colony, back towards the jeep.

'Elizabeth,' he called, 'before you go back to the blind, come over here. Let me show you this.'

She walked over to him. 'And by the way, I'm Alan.'

'Fine,' she smiled.

They stopped amidst the grass about a hundred and fifty metres from the hide, and the scientist stooped and began to pick up small bones and skulls.

'This is a killing field,' he explained. 'These are the bleached white bones of seals, remnants of long-past seal harvests when juvenile males were herded up here and then clubbed to death for the fur trade.'

He picked up a skull no larger than his own clenched fist and reflected quietly.

'Profits from the sale of fur seal pelts paid the purchase price of Alaska within ten years of its acquisition from the Russians. That was in the middle of the last century, but with slaughter on such a vast scale the fur seal herd numbers plummeted. So in 1911 the major powers got together and signed a treaty, and numbers recovered. A lot of seals have been slaughtered since then, but in 1972 the killing stopped altogether. This latest decline, which seemed to begin in the late fifties, has continued despite the fact that seals are no longer killed, and we don't know why. It's a mystery.'

'So it wasn't the harvest?'

'No. That's one thing we are sure of. It wasn't the seal harvest.'

Slowly and with much apparent thought, he laid the skull down and stood up, gazing inland towards the green cliff face which overlooked East Reef Rookery.

'No, whatever it is, it isn't just seals. See over there, on the green cliffs? Ten million birds come back every year, rear chicks and feed off the rich waters of the Bering Sea. They've declined, too.'

'And you have no idea why?'

He shook his head slowly, and then dramatically but with studied solemnity, he walked away, leaving Elizabeth to ponder. In the distance a curious group of juvenile male seals had gathered together near a large tree trunk washed up by the violent storms of winter. Their heads wavered and wove, bobbed up and down as they looked over the debris and tried to determine what the humans were doing. Dr Steinberg increased his speed and strode purposefully to the jeep, while Elizabeth returned to the hide to continue her research observations and the juvenile seals settled down to sleep again on the grassy field which had once seen the death of a million of their ancestors.

The powerful binoculars swept the beach and the trained, observant eye picked out the birth of one more pup and the return of two more females. Elizabeth recorded it all according to the yellow numbered squares. A raven flapped down briefly and a pair of rock sandpipers jumped waves at the water's edge. Farther out, a line of least auklets flapped low along the coast and out to sea. Not much bigger than a sparrow, with dark backs and pale bellies, they were flying out to feed on plankton while Amik and her mother continued their urgent cycle of feeding and rest.

In the middle of the day, when most of the fur seals were asleep or at least soundly resting, there was a soft cough at the western edge of the rookery. The fox stood on a prominent boulder and searched the beach with an intent and hungry gaze. The two rock sandpipers were still poking about between pebbles with their long beaks, looking for shellfish or sea worms. The fox loped along as if she had not seen them, then suddenly stopped about forty metres from the birds, held her tail low, almost between her legs, and, crouching, nose close to the ground, meandered on tip-toe across the beach in the general direction of the feeding sandpipers. The birds seemed unconscious of the fox and continued with their pecking and turning, concerned to avoid the odd seventh wave which broke and flooded farther up the beach, almost wetting their feet. A resting male seal rolled a watchful eye in the general direction of the stalking fox, but neither moved nor barked.

A distant raven called from the interior of the island and

another big wave broke, and as it did so the fox leapt. Too late! The birds had already seen her and flew off, rattling annoyance and calling shrilly. Somewhere over the coastal plain, well hidden in the tussock grass, were their nests, each holding four young chicks. The fox stood stock still and looked up, watching the fluttering of their wings and following the line of their flight, but the birds were astute enough to know that to fly directly back to their nests would have led the clever opportunist to another meal. They circled the beach, calling, watching and waiting, still hungry for food, and when the fox had moved eastwards they returned to the beach, this time a little farther to the west, well out of the way of the hunter. She, however, had already forgotten the sandpipers and was moving rapidly, nimbly, across the black boulders towards the female carcass.

# 3 IN THE BEGINNING: AN OLD, OLD TALE

'Many, many years ago, when history was told in song instead of in books—' Alex looked directly at Elizabeth when he said this—'there was an Aleut chief on Unimak Island who had a son called Igadik, who spent much of his time in his kayak hunting the whales, otters and fur seals which roamed the waters around the island where he lived.'

Elizabeth looked across the room and out of the window. From her position at the table she could see directly out to sea. There it was, the Bering Sea, cold, grey but teeming with life.

'Igadik was a highly skilled navigator, as well as a hunter. He it was, of all his tribe, who could spend days in the open ocean away from the sight of land and never get lost.'

Breakfast was over. The sun was high and Alan wanted to get on with the day's research work, but Alex had popped in during breakfast and was telling yet another of his Aleut stories.

'Each spring Igadik watched the fur seals swim north through the Unimak Pass, heavy with unborn pups, and every fall they swam back, going south. Yet, although he and all his ancestors had searched the waters north of Unimak, no one had ever discovered that cold shore where the fur seals rested and gave birth. Then one clear day, after Igadik had put to sea, a fierce wind blew up out of the south. It was so strong, even Igadik with all his skills was forced to run before that wind. For many hours he struggled and worked to keep his craft afloat. Finally the storm exhausted itself and Igadik, tired out by all the stress and effort, found himself in a deep, swirling fog.'

Elizabeth admired his charm. Here he was, imitating an Aleut shaman telling a long-told tale, proud of his own people's culture.

'Through the fog Igadik could hear familiar sounds, almost as if he was close to his own home: the cries of the birds on their cliffs and seals, calling. The voices came from a land hidden in the mist. He worked his tiny craft forward, through the swirling clouds, until in front of him loomed up a towering cliff, a dark coastline. Cautiously he paddled slowly, parallel to the shore, and to his utter amazement he saw millions of fur seals, crowded so close together they almost seemed to be lying one on top of the other, and most of them were nursing pups. There were so many seals on the black beaches and choking the waters close to the shore that he had to search for a long time to find a place to land.'

'Come on, Alex,' urged Dr Alan. 'Finish the story.' He sounded quite impatient. 'Elizabeth's got to get out to the blind.' Alex nodded to acknowledge he had heard, raised his eyebrows, smiled and carried on.

'Igadik spent a year and a day living on that misty island, gathering fur seal skins to take home to his people. On very clear days he could see another island to the south.' He looked up and smiled at Doctor Alan. 'Won't be long now, Doc, but must just finish the tale, eh?' He fiddled with the maple syrup jar, looked to Elizabeth for reassurance and continued. 'It was a damp, cold place, that island, and the young man longed to return to his own hearth and home. Sure, in the summer, when the sun, very occasionally, came out, it was pretty—covered with flowers of brilliant colours—and in the winter the wind blew the mists away, leaving the snowy volcanic island sparkling in the icy sea.'

The word 'volcanic' jarred in Elizabeth's hearing. An old Aleut storyteller would not know that, say that. This was Alex saying he knew more than his ancestors, asserting his equality.

'Then came that day in summer when a very strong north wind blew. Immediately Igadik loaded his kayak with seal skins and set his course due south, running before the wind, until he sighted the familiar volcanoes of Unimak and the land of his father. The people of the village could hardly believe their eyes. This their son was lost and was now found. So that evening many songs were sung and many dances danced and Igadik told many tales of the misty fur seal island he called Amik. In years to come those songs were sung to Unungan children. But no one ever retraced the steps of Igadik. Only his fame and his songs remained.'

Alex paused for effect. Doctor Alan mistook it for the end of the story and stood up, but Alex continued:

'Then, after many, many hundreds of moons had waxed and waned, strangers came from the west and spent years searching for the seal islands. They wanted to kill the fur seals, wanted to skin them for their pelts, to sell them; but the descendants of Igadik said not a word. Eventually one man found the islands which Igadik had already named. But because that man, Gerasim Pribilof, did not know of Igadik and his story, he named the islands after himself. And then began the killing of the fur seals.'

There was a moment's silence, then Elizabeth said:

'Did you know that story, Dr Steinberg?'

'Yeh. I think I've heard Alex tell it.'

Alex chuckled. 'I should think so, Liz. Dr Alan's been here long enough to hear my stories at least five times apiece.'

Alex Merculieff was more than an Aleut assistant. During the summer he was a friend and a confidant, a link between Doctor Steinberg and the island community. Alan moved across the room.

'Look, you guys, can we get on now?' He looked directly at Elizabeth.

'Sure, sure,' agreed Alex. 'Out in five minutes, Capt'n,' and he gave a mock salute.

* * *

A lone black-furred pup was bleating feebly as it crawled very weakly, right at the edge of the beach. All night it had searched out milk from nursing seal cows, but to no avail. It had been snapped at, bitten, pushed roughly away and was now starving without any hope of receiving any sustenance at all. In its intestines, tiny adult nematodes guzzled greedily on its blood. The first the pup saw of the fox was a small sharp muzzle bobbing up from behind a black rock. The light-footed creature leapt nimbly and stood stock still. Its penetrating stare scanned the beach. The pup was alone. It had crawled out of any defended fur seal territory and stood wearily, its ears standing comically proud of its head and its large wet eyes rolling, unfocused. It waved its head feebly and bleated incessantly. Somewhere out in the Bering Sea a female fur seal had died or been caught up in a fishing net, or had simply given up the struggle to raise a pup. The abandoned offspring gave a weak groan and lay down, short, sharp breaths lifting its abdomen. The fox looked round. Stepped carefully, all four feet in sequence, down the rocks which lay between it and the dying pup. The predator moved forward. The pup turned its head to face the source of soft padding feet, exposed its throat and the fox killed it. There was hardly a struggle, just a few muscle spasms, and then the dead pup was dragged off the beach onto the grass and under the weathered roots of a beached pine tree trunk.

The fox was about to tear at the still warm flesh with its sharp incisors, when the noise of Elizabeth's tribike disturbed it. Out of pure enjoyment, she hurtled down the rough track, bumping and skidding, revelling in the sense of freedom the island gave her when she was on her own. She sped past the tree trunks, did not see the pup carcass or the small head and sharp brown eyes which kept her in tight focus as she slid to a halt and switched off the engine. While she parked the bike and walked to the hide, the fox dragged the corpse into the lee of the rocks and began tearing at the meat.

* * *

July 12th. The sun had nearly completed its low arc across the southern sky. It was the evening of Amik's fifth day, and on mist-covered St George light was reduced to a glimmer. Males moved across wet rocks, shaking their wet manes. It had rained most of the day and the pup's black coat was a series of irregular, sodden clumps of fur. She was asleep, lying close to her slumbering mother. Both creatures breathed heavily, deeply, in and out, oblivious of the activity around them. Female seals were coming and going, crossing the beach and swimming out to sea, or returning and calling, responding to the bleats of their hungry pups. Hurrying, dodging, trying to evade the unwelcome attentions of harassing territorial bulls.

A large black shadow alighted on a rock close to the sea's

edge, and croaked. The feathers on the crown and shaggy throat of the raven rose up. Its stout bill was wide open. The pup shivered and started. Looked up. The bird fed on carrion and was looking for debris. Sixty-three centimetres tall and standing dimly silhouetted against the dark sky, it looked twice that height to the small pup. Already she had seen this male bird flying over the beach, its heavy head, extended neck and diamond-shaped tail distinguishing it from all the other birds of the shore. Even the black shapes of cormorants which occasionally flew by, heavy and low, over the waves just offshore, could not be confused with the raven. It croaked again. The deep crow-like call rasped in the small ears of the pup. It was another sound she had learnt to associate with the beach, but not so close.

Now the bird was jumping down onto a lower rock, walking on stiff legs towards her, its beady brown eyes with their intensely black pupils darting glances left and right and down, straight at the pup whose brown eyes were wide open and wet, staring up at the bird. She turned to look for a teat. Her mother moved slightly to make herself more available and the pup smelt a different smell, rank, full of meaning. The bird took in its flavour, too, and paused. It turned its head sharply and suddenly, twisting a glance over its shoulder, and saw the lumbering male moving over the rocks, checking the breath of each female. The raven called again. The male seal reared up, bared its teeth, shook its heavy, sodden mane and barked.

In the hide Elizabeth had noticed the raven, too. The bird had attracted her attention to grid square 14, and although she could see quite well (the square was no more than twenty metres from her observation post), she picked up her binoculars and focused on the female seal as she stood up and rolled over slightly so that she was sitting, resting on her right front flipper, with her pup nibbling, very satisfied, at her teat. Unknown to Elizabeth, the new scent, that different smell, was full in the pup's nostrils. The male smelt it too, deep in his consciousness. He hurried across the rocks. The raven backed off, flapped up onto a wet rock, the mist clinging to its feathers, paused, and then slowly, with much labour, took to the grey air with noisy flaps of its heavy wings, moving east, farther down the rookery. By now the male was at the female's side and he barked softly. The female snapped back, he barked again. The female moved forward and the male nuzzled her neck, briefly, took in the full scent of her breath.

The female seal was on heat. Elizabeth recognised it from the way the male was behaving and the way Amik's mother was letting the bull approach her. Doctor Alan had explained to her the difference in interaction between the male and the female, at this particular moment. According to his research findings, females co-existed with males on the beach, but hardly tolerated them apart from this one, significant act. He had told Elizabeth that it was the presence of a fertile male which brought the female fur seal into oestrus and that she would mate once only, be impregnated and then go out of oestrus immediately, unless the male who covered her was immature or non-territorial; then she would remain available until a territorial bull seized her, provided that was prompt, because, according to previous research findings, females remained on heat for no more than thirty-six hours. Such was the efficiency of the evolutionary programme.

The lenses of the scientist's binoculars glinted dimly in the weak light. For a moment the pup's attention was attracted to the open window of the hide and Elizabeth's presence. But it was that rank smell, so unlike anything else she had ever smelt, that dominated her mind, and the male so close and being tolerated for so long. The pup's mother moved two flipper-lengths up the beach, away from the pup, and the male lumbered in; touched her mother. So close! The pup retreated from the overwhelming shadow, cowered against a low wet rock, and watched. The other females sat up, their heads waving madly on their long, sinuous necks, barked at each other, snapped, glared down at the female lying close to a large black boulder. The male, as his nostrils traced the outline of her face and neck and sniffed at her length, edged closer, awkward on the uneven rocks, but closer still. There was no courting, no bond. He was a rampant bull fur seal and she was a cow on heat. The enormity of his full weight covered the female until she was just a small brown head breathing heavily under his judderings. By the end of the night her breath would be sweet again. She had mated here before, close to that very boulder, but not with this male. Last year it had been a different bull. The large male fur seal who covered her now was not Amik's father.

So Amik's mother was impregnated, the next stage in the inevitable programme which all fur seals followed. The bull lumbered back to his vantage point high above his territory, lay down, closed one eye and dozed. The pup returned to the teat. The female, was edgy. She turned; turned again. Sat up. Lifted her back flipper. Scratched her neck. Looked down at the small pup. Slowly closed her eyes and almost imperceptibly, without disturbing her pup, lowered herself until she was in her resting position once more; almost asleep on the cold damp shore of St George.

In the distance there was the sound of a tribike starting up. Elizabeth was leaving her hide, returning to the field station for the night. Already the female was off heat and Amik, well fed, was resting too. The smell had gone, was almost forgotten. Farther down the beach the raven was pulling at a piece of carrion it had found deep in a crevice between two smooth-sided rocks. Its long bill was poking, tearing, its head nodding violently to give more power to its efforts to tear at the small blubber-bloated carcass. Darkness was slowly enveloping the scene. The bird pulled back, turned, and with wide beats of its broad wings fell into the wet breeze, let the Arctic wind carry it inshore, away from the seals, back to the high nest on the stark cliffs which reared up from the plain.

The short night fell. Out at sea a right whale blew. Another tired female seal swam back, heading towards East Reef Rookery, past the skin plant, past the killing fields. In the darkness she slid and fumbled up out of the water onto the hardness of the cold shore. A bull seal barked in the darkness. There was confusion, noise, pebbles sent scudding. A small boulder was overturned, tumbled noisily down into a crevice, a pup was torn from its mother's teat. The female panted up to the gaggle of five other waiting females in a territory four down from Amik's. The rookery quietened. The males lay down. The pups sucked, the females dozed and the mist changed into the regular patter of hard rain on the rocks. Out along the shoreline the tired waves lapped gently, hardly breaking.

Inside Amik's mother the evolutionary clock jerked forward. Her programme now dictated that she must go to sea, feed, build up her waning strength and thus produce more milk.

'Energy transfer to pups comprises two thirds of the female energy budget on land, but only twenty to twenty-five per cent of energy consumed at sea goes to storage purposes.' Doctor Alan was giving his nightly lecture. Elizabeth was listening intently, for she had so much to learn. 'The high rate of transfer to the pup on land and the low rate of transfer to storage at sea might seem to be inconsistent,' he continued, 'it might seem to suggest a substantial net loss to the female; but the difference is partially explained by the fact that the time females spend at sea is much longer than time spent on shore. And a lactating female eats over one-and-a-half times as much as other adult females. So although percentages might seem small, the gross total is much larger.'

'Is that so?' interrupted Alex. They were all sitting around the table eating halibut, fresh caught by Alex and donated in a huge slab of flesh that hardly fitted into the fridge. Alan had steamed some of it in milk. It was delicious. Was there anything this guy could not do? thought Elizabeth.

All that short night, in the teeming rain, the female fidgeted. The pup tasted the water in her milk: fat content was noticeably lower now, the richness of birth had dissipated. The time had come when the female had to return to the sea.

Another cold, wet day dawned. Six kilometres away from the rookery, the scientist emerged from his field station wrapped up in warm clothes and oilskins. He wiped the rain off the seat of his tribike, mounted, turned the ignition and had trouble in starting the machine. Time and again he pressed the ignition button. Hopeless: the dampness had got to the leads. In his hide was a long pole, two and a half metres long with a noose at the end. The scientist remembered it. Remembered also the red straps and the tiny computer in a small canvas bag on the sideboard in his house. He sighed, frowned, released the parking brake and wheeled the bike down the hill into the skin plant, where he gave the superficial leads a cursory wipe and left the machine to dry out.

The plant echoed with memories of the days when two hundred Aleuts would bring the slaughtered remnants of the seals, pelts, into the huge building, soak the skins in brine and prepare the fur for sale on the world market. Here, Alex's father had crushed his hand one frantic summer thirty years ago, when the harvest was big and the overseer, keen for production, had pressurised the Aleut into a mistake. Alex and he had found an old accident book when they were browsing through a pile of discarded Seal Company papers. Alex had come up with the idea of an island museum. He had been trying to persuade the islanders to hunt for mementoes. Now that the seal kill was finished the young Aleut wanted to preserve something for posterity, to record all that effort, all those years. Alex had snatched the book from Alan and turned the pages until he had found it: 'Vladimar Merculieff injured left hand in pressure rollers.' One terse sentence.

The doctor glanced across the plant. Now it looked like a garage workshop, with outboard motors, a couple of all-surface vehicles, and a very old and battered jeep, occupying spaces between what used to be brine vats and presses.

The door of the skin plant creaked open. Elizabeth walked in. She had come to collect her tribike. Sensible woman, she always parked it under shelter. They were supposed to be all-weather machines but the Pribilof damp seemed to get into everything.

'Hi!' she smiled.

Doctor Steinberg nodded back.

'I'm a bit early this morning, thought I'd get out to the rookery ready for you and Alex.'

'Fine. Lock up the plant when you've finished,' he ordered, throwing the keys across. 'Brought the other bike down here. Got wet. Won't start. I'll get the jeep, pick up Alex and meet you out there in an hour.'

'OK,' smiled Elizabeth.

Alan walked back up the hill to his jeep. He checked the back of the pick-up. There was the 'guillotine', a stout wooden device which looked just like a medieval stocks for a one-legged man. He pushed at it, checked it was secure, then walked back into the timber-planked cottage, picked up the canvas bag which held a tiny computer and red straps, and returned to the jeep. He got in and drove off to the village, which was officially called the City of St George.

On the beach the bulls were doing the rounds of female breath while juvenile males sat on rocks in the breakers grooming, watching, waiting for the chance to battle for a territory. The tribike had already arrived and parked. Elizabeth was busy on her count of newly arrived females and newly born pups when the noise of four diesel-driven pistons and four heavy wheels came bumping across the grass. It frightened the young males who slept soundly on the hauling grounds—the killing fields.

The juveniles waddled and hurried back towards the beach, fleeing the squealing brakes. Amik did not see the jeep but heard voices, the banging and the unusual movements, sounds of lifting and dragging, She looked round, clambered up onto a higher rock to get a better view. As she did so, her mother moved off, started to leave the territory. The bull roared, lunged across at the female. She snarled back and fled. The pup stood still, watching her mother hurrying away, from the bull, from her! Down the beach, towards the sea.

A slight noise higher up the beach distracted the pup and she turned to look. From her vantage point she could see, twenty metres away from her, two heads with bobble hats: men, crouching, creeping, coming towards her, both bearing long stout sticks. Suddenly they were standing, running, quietly determined, working together as a team. The man with the noose was chasing the pup's mother. The man with the long stout pole was fending off the angry bull seal. He placed the pole across the creature's neck, trying to pin him to the ground, avoiding at all costs the sharp yellow canine teeth and the threatening lunges. Behind the man another bull on another territory was snarling at his back, while under his stick the resident bull was heaving powerful shoulders. Neither man spoke. Both knew what to do. They had rehearsed it, carried out the same movements many times before.

The scientist was running across the rocks, agile, skilful. He aimed his long pole, thrust it out in front of him. Round the female's neck went the noose. Doctor Steinberg turned the stick, turned it again. The rope tightened, took a grip, and he hurried backwards, dragging the female with him, bumping over the rocks, up the beach and out of the possessive reach of the bull; away from any occupied territory, back to where the two men had unloaded the guillotine. Immediately Alex released the bull and rushed back to the guillotine too. He watched as the scientist deftly inserted the head of the seal into the wooden slot. Alex closed the jaws of the guillotine. Held by the neck, the female lay quiveringly still and Alex picked up a large pebble and adjusted it under the head of the seal, making it almost a pillow on which to rest her head. The scientist reached for his canvas bag.

He was paid to explain the decline in the size of the fur seal herd. The reason was all-important because it was more than the loss of one species, the fur seal: it was the total decline of all species in the North Pacific that worried his agency. After years of research there seemed to be only one conclusion: evidence for the cause of the decline did not exist on the land. Hence the capture. The scientist had to search for a reason in the deep ocean. So on that bleak shore, early on that cold summer day, he had captured Amik's mother and, while the pup looked on helpless, the two men strapped a TDR (time, depth recorder) in position. In ninety seconds the scientist was crossing his fingers and his Aleut friend was releasing the seal, lifting the wooden jaws of the guillotine and poking the seal back into life.

Bemused, she stood up. Around her chest the bright red strap held firmly but not tightly, binding the research computer to her back. She was disorientated for a moment and waddled unsteadily up the beach. The Aleut saw this, moved quietly and efficiently, and stood in front of her to block her progress. She saw him, turned, saw the wide, long shivers of water in front of her and tumbled down the rocks, fighting off the males, and slid back into her true element, the ocean. The men smiled, scratched a forehead, loaded the guillotine back onto the truck.

The female seal had moved out beyond the sounds of stridulating crabs at the sea shore, beyond the wash of the waves against the storm-tossed rocks; now bubbles raced past her ears, the pressure of the water squeezing out air from her fur. She was seven metres down, pushing herself through the kelp where the shrimps clacked their claws. Deeper she swam, and the dense chatterings sizzled. Somewhere she heard a fish sing—the sound made by the swim bladder, the buoyancy chamber most bony fish possess, which doubles up as a resonance chamber. There was food here, fish for the taking, and she was hungry. In the kelp forest ahead of her she saw the flash of white scales, and like a banshee of the deep herself, she whistled weirdly into her hunting routine. Meanwhile, on shore her pup was alone, standing on her front flippers and looking out to sea.

Silver light glinted across the low surface of the sea. The trio of scientist, assistant and Aleut stood looking out to sea, watching to see if they could catch sight of the female fur seal, Amik's mother, leaving the island to hunt, carrying the TDR with her.

'That water's cold out there,' murmured Alan. 'Three minutes, that's all the Aleuts give a man who falls in. That's right, isn't it, Alex?'

Alex nodded.

Elizabeth believed them. Even on shore, in the summer sun, it was cold.

# 4 RESCUE

'Sure, it'd cut across the flight path of the least auklets returning to roost and the red-legged kittiwakes picking up mud for their nests, but for goodness sake, how many planes'll be landing each day? Better to keep all human activity to the same side of the island and close to where humans have always been, than putting the runway on the south side. There's enough disturbance there with all the harbour works. The fur seals on Zapadni aren't so used to humans.'

Elizabeth and Dr Steinberg were talking about human interference in the Bering Sea and the plans for the island. It was quite clear he did not like the harbour works on the south side of St George Island, and the thought of a runway there as well worried him. However, wherever the new runway was built it would cause disturbance. He knew that. Slowly Elizabeth was beginning to realise how much effect humans were having on the marine life in what she regarded as a remote and isolated ocean.

Nearly 2,000 tons of fishing junk were released into the Bering Sea and North Pacific Ocean every year. Plastic fishing nets, plastic lines and plastic containers had replaced the natural fibres such as hemp, cotton and linen. These new materials were cheaper, lighter and stronger, but plastic was very resistant to weathering and floated about for years and years, a danger to the sea creatures which swam through it. It was into such a contaminated ocean that the female swam, leaving Amik alone on the cold shore.

Every time Amik's mother went to sea to hunt, she undertook a round journey of four hundred kilometres to her feeding grounds. Once she was out, beyond the unwelcome attentions of love-hungry juvenile male seals who crowded the inner one hundred-metre band of coastal water, she porpoised for another hundred metres and then settled to an efficient pace, hunting as she went. She dived to depths of over a hundred metres, her heart-beat slowing from one hundred and ten beats a minute at the surface, until it was a fifth of that. Although a comparatively shallow diver as far as seals are concerned, the female still had to contend with the problem known to human divers as 'the bends'—decompression sickness.

The bends are caused by supersaturation of the bloodstream with nitrogen at pressures of several atmospheres. When the pressure is released by the seal returning to the surface, the gas forms bubbles, just like champagne when the cork is released. It is an agonising and sometimes fatal condition. By exhaling before her dive the seal limited the amount of nitrogen available for absorption into her blood-stream. This effectively halted nitrogen absorption; gas that remained in the lungs was then compressed into parts of the respiratory tract which did not exchange gas. Nitrogen also has a narcotic effect under pressure, inducing a state known as 'rapture of the deep', similar to drunkenness. But evolution dealt with all those problems. When Amik's mother dived, the slits of her nostrils closed automatically under the pressure of water, and the soft palate and tongue together at the back of her mouth closed up, so that when she needed to open her mouth to catch prey, water was not forced down into her oesophagus and trachea.

She was not a true seal: northern fur seals are eared seals, related very closely to sea lions. It is true that seals like the Weddell and Elephant can dive much deeper and for much longer periods than fur seals, but Amik's mother shared some physiological characteristics with them. For instance, she had a greater blood volume per unit of body weight than most other mammals, and that blood contained much more oxygen-carrying haemoglobin than man's. Her muscles were adapted, too. Myoglobin concentrations, which help bind oxygen, were much greater in her. So she was able to store oxygen supplies when she dived and, to ensure the efficient use of that oxygen when she swam underwater, her blood supply was diverted largely to the brain. In this way she was able to dive with little air in her lungs, whereas man, to increase his breath-holding capacity, would have to take in more air. Because of these physiological features, the fur seal was programmed to deal automatically with deep sea diving problems: all she had to do was find the fish, while her bodily responses coped with the environmental changes.

On shore, while her mother searched for fish, the pup needed to find comfort, warmth and protection. She slid down the rock, with its yellow-painted line, struggling hard up the next, over it, and down into another crevice; up again, over the next and down close to the sea. There, in the lee of a very large boulder, were several rocks of a similar size, all worn smooth by the waves washing back behind that large boulder. It was a small platform, and ten other pups had already congregated close together—all without their mothers, all waiting.

Amik saw them. She struggled over more rocks, skirted the barking, snapping attentions of a group of females on grid square 10, and began to struggle up onto the platform. The other pups, writhing, wriggling, continually moving about, hardly noticed her. She pushed hard on her back flippers, her front flippers sliding across the wetness of the rock. One of the other pups peered curiously over the edge of the platform, too close for Amik's liking, and she snapped hard at its neck, but in doing so she slid back down to the bottom. There she steadied herself, edged sideways along the base of the platform and tried again. This time another pup greeted her with an open mouth, small sharp teeth showing. Amik ignored it, pushed as powerfully as she could with all four flippers, and lifted herself up onto the platform, shouldering the other pup out of the way. Taking the force of its bite on her well furred shoulder, she rolled into the centre of the group. There was a flurry of flippers. She looked around her, snapped at two pups, jostled a third and established herself away from the edge of the platform, close to the shadow of the rock and protection from the wind. She shook herself, looked around, then lay down, curled up and rested.

Out at sea a Japanese trawler, two days out of Dutch Harbour (the large fishing port on the Aleutian Chain of islands, 500 kilometres south of St George), had found good fishing. Underwater the noise of bubble cavitation from the propeller blade's churnings, and the roar of the nets being dropped over the side, sent great walls of sound through the water, overwhelming creatures close to the boat. On board, one hundred tons of pollock were trawled up in a single net and disgorged onto the deck, where the process of treatment and storage began immediately.

The female had swum hard and fast, following urgent lines of birds streaming low across the flat waters of the calm, dark sea. It was a diving murre that first attracted her attention. She was about to attempt chase when it surfaced. She followed and it took off and she saw more birds in the distance, flying away from her. Birds always saw more than she did. They worked the waters, too. They fished. She often watched for them, used their eyes in her hunting. That night she did as she often did and followed them. Porpoising from time to time to gain speed, she hurried forward through the near-calm waters, the whispered beats of a thousand wings above her head and the speed of flight urging her on.

Like an enormous magnet the fishing vessel attracted thousands of seabirds. The fish oil jettisoned by the boat could be smelt and savoured by storm petrels, gulls, even murres, from a distance of over five kilometres. The birds flew low and fast in their hundreds, swarming all around the boat, and fur seals, picking up clues from the hunting birds, joined in. One of those seals was Amik's mother. It was not only fish debris that the fishermen threw overboard: discarded pieces of net floated side by side with mats of kelp. In the darkness it was difficult to distinguish between the two.

She was a solitary hunter. She had no intention of sharing her hunting waters with any other fur seal, but the waters around the Japanese boat were rich. So, on that cloudy, moonless night the female found herself in a loose-knit hunting group. Around the trawler there were several other females and two immature males. She ignored them, simply tolerating their presence and they hers. There was no interaction, it was as if none of the creatures could see each other. They fished independently, and always fed primarily at night because the important food species rose to the upper layers during the hours of darkness. Squid were their mainstay and small schooling fish, such as sand lance, capelin (a small fish resembling a smelt), and herring supplemented the diet, as did the wall-eye pollock.

She paused a hundred metres from the boat, heard the noise, saw the lights and took stock of the situation from a position slightly hidden under a mat of kelp. The trawl net had been drawn up and much debris had been cast overboard. The crew was now preparing to send the trawl back over the side. All the seals watched the ghostly lights of the vessel from a safe distance, hiding in kelp mats, peeping from time to time through the vegetation, watching warily, then plunging forward and swimming down, turning to surface and watch again. The food was plentiful and the seals ate well.

Suddenly the engines of the boat revved up. There was a great clanking of pulleys and the net crashed back into the water. A fur seal thrust its head up to get a better view and for protection pushed forward into a mass of discarded nylon net, mistaking it for kelp. Instead of the soft sliding greenness slithering over its coat and away into the distance of the dive, the blue strands clung close, held tight to the neck, became entangled in the fur and settled down into the pelt.

* * *

It had been another short night—cloudy, dark, cold and damp. Amik had lain close to the other ten pups on that narrow sheltered platform, listening to the lapping of the quiet waves and the recognised sounds but unseen activity of the beach. As the darkness faded she heard a familiar cough, and scurryings on the higher rocks a little away to the east. In the lightening gloom she saw them: a pair of foxes on patrol, looking down at the pod of pups, most of them still curled up asleep. The two scavengers stood motionless, but for intense searching stares and slight movements of the neck and head.

Amik jumped to her front flippers, raised herself high, opened her mouth and bleated furiously. The foxes ignored her but her urgent movements and shrill cries awakened the rest of the group. All the seal pups fidgeted up and stood with querulous eyes, gazing at the foxes from a phalanx of bleating immaturity. They stood close together, waving their heads on unsteady necks. One fox disappeared down into a crevice and came up on the other side of a large rock, closer to the pod. The other fox moved along the beach, pausing closer to the pups, too.

This was too much for Amik: she wobbled forward, over another pup, which snapped at her. She took no notice, reared up onto another pup's back and bleated long and hard. The other pup squirmed and wriggled forward, trying to dislodge Amik who finally relented and slid back onto the rock platform, but without taking her eyes off the foxes. Two rock sandpipers burrowed and pecked at the sea's edge as if nothing was happening at all. Suddenly one fox rushed forward, straight at the pod. Some of the pups flinched and moved back, but Amik held her ground, barked a high-pitched bleat. The sandpipers flew off, a single call of alarm thinly echoed over the shore, and the second fox watched as the first ambled nimbly up and down rocks in the general direction of a female seal skeleton which still lay on the beach, although most of the flesh had already been eaten.

Elizabeth was sitting in the cab of the jeep. She was feeling cold and miserable. Here she was on a beach in July, but in a parka and thick gloves—what a way to spend the summer, she thought, perished to the bone! She was bouncing about in the jeep with Doctor Alan at the wheel, travelling along the uneven, muddy trackway. Years ago, before the advent of four-wheel drive and internal combustion engine, these trackways had been lined with wooden planks to let the horse-drawn carts bring the sealskins down to the processing plant, built close to the small harbour. Now that there was no seal harvest, traffic on the trackway to the rookery was very light.

Alex was sitting in the back of the pick-up, perched amidst capture gear and holding a canvas bag containing another TDR. It had all been agreed at the nightly conference after dinner: fix another TDR. Alan parked the jeep close to the catwalk and got out. Elizabeth let Alex unload what gear there was and, walking cautiously towards the edge of the grass, took out her binoculars and scanned the rookery. She saw what she took to be a female landing a hundred yards down to the east, in the no-seal's-land which lay between the two rookeries of East Reef and East Cliff.

'See it?' She asked Alan who had walked down to stand at her side.

'What?'

'There. A female come ashore?'

Doctor Alan did not reply. Instead he tried to locate the creature through his own binoculars.

'Emmmmm. No. It's not a female. Juvenile male, I'd say. Coming onshore to get to the hauling grounds.'

'What's that blue thing?' she asked. 'Sort of a line round its neck.'

'Oh.' Doctor Alan groaned. 'Plastic net, I guess. We see a lot of 'em.' He turned and shouted, 'Alex!' and walked briskly away from Elizabeth, back towards the jeep.

Alex had already unloaded the guillotine and was carrying the two long poles with nooses at their ends towards the rookery.

'Problem, Doc?' he frowned.

'Young male out towards East Cliff. Net entanglement. Want to have a go?'

'Sure. You bet, Doc.' He sounded eager.

The young male seal had struggled up the beach and lain down on the grass amidst a group of other juveniles. There he rested, the blue nylon net clinging tightly like some ugly, choking necklace. All around the resting and sleeping seals were the bleached trunks of trees, forest giants that had survived a storm-tossed journey across the ocean, either from Alaska or Siberia. They had been washed up on St George's shore, thrown high above the rocks of the beach as graphic evidence of the intensity and rigours of spring storms.

If any of the sleeping seals had been awake and lifted its head to look over the twisted and gnarled roots of one storm-abandoned, white-washed birch, it would have seen two stealthy heads creeping with intent, approaching the hauling ground from the beach.

Suddenly Alan shouted: 'Now!'

Alex jumped up and started roaring, banging a piece of wood against the long pole he carried in his right hand. Both men charged head-long, one either side of the huge tree trunk; Alex driving, Alan with quieter footwork, holding tight to his long noosed pole, keeping very close to the tree trunk. The seals were up and running, as fast as the men, away from the sea and onto the unfamiliar territory of the untrodden tussock grass. There were a hundred of them wheeling in an uncertain mass, sleek, dark-grey bodies and, in the middle of the group, one with a hideous plastic necktie.

'Now,' urged Alan, and Alex lunged forward and, with a huge thud of his pole against the ground, divided the seals in half, making the entangled seal the leader of a second group. In those moments of hesitation, while the halted seals turned to regroup, Alan was in there with Alex who cut off the retreat of the injured seal. The scientist ran forward with his pole, aimed, and had his noose round the young seal which pulled violently and barked fiercely. All the other seals had veered away again, and were rushing back down towards the sea. Alex also looped his noose over the seal's head. The men were pulled this way and that as they struggled to tighten their grip. The seal curved its back, reared up, twisted, turned, lunged at Alex who fell onto one knee with the force of the seal's strength but was back on his feet in an instant. He turned his noose tighter and forced the creature down onto its face. It squirmed, wriggled, turned its head biting the air, barking vehemently. And then for a sudden moment the seal was motionless. Elizabeth held her breath.

'Right,' whispered Alan hoarsely. 'Hold on in there, Alex.'

He dropped his pole, threw aside a glove, reached into his pocket, retrieved his knife, knelt down at the side of the seal, well back from the head, and began to cut the net away. It was thoroughly embedded; the wound was red and raw, deep into the flesh.

'At least it's not infected,' he whispered. 'Come on, old fellah. Nearly done.'

The last thread was cut, and then Alan had to pull the net out from where it was loosely trapped, deep inside the cut. He stood up, folded his knife shut and pushed the net fragment into his pocket before putting his glove back on. The seal trembled, flicked its tail violently. Alan jumped back to his pole and clasped it firmly. Neither man said anything as they let the seal stand up. Alex slowly unwound his noose and pulled it over the head of the seal, then stood back, ready with his stout pole to give Alan a hand if necessary. Alan flicked his wrists, untwisted the rope, and pulled the pole away. The seal snarled at him, turned to its right, lunged briefly at Alex, but almost as part of the same movement was running at top speed down the beach.

Alan went across to Alex and put his arm around him. Alex slapped Alan's back in response. Both breathed heavily.

Elizabeth hurried across and all three stood and watched the juvenile tumble out of sight, down from the field onto the rocks of the beach. It was roars and scudding pebbles all the way as the young male blundered straight across the rookery. Amik saw it go diagonally from east of the hide to close to the rock platform where she huddled with the other pups. At least one territorial male made an unsuccessful pass at the fleeing creature, but it ran so fast, took the whole beach by surprise, that the only injury the seal carried into the ocean was that deep, unhealed scar which completely encircled its lower neck.

* * *

Out at sea Amik's mother was diving again. She breathed very deeply for several moments before she dived; by doing this she increased the oxygen supply in her bloodstream before exhaling the air from her lungs and then sinking below the waters. The rolling, scraping, drag of the trawl net along the ocean floor masked all other sounds. Each time she dived the TDR strapped to her back recorded essential information. She had been underwater for three minutes and was following the net at a depth of seventy-five metres. Her fur, with more than 100,000 hairs per square centimetre, was so impermeable to water that her skin remained dry despite the pressure of the deep sea. Her nostrils were tight shut and her small external ears, tightly rolled cylinders with a narrow waxy orifice, prevented the entrance of any water. It was dark. She moved forward, her whiskers trembling, feeling, and suddenly she lunged and snaffled. A pollock held by her thirty-six teeth had no chance of escape. The lower incisors fitted snugly into a notch in her upper teeth, and the upper molar and pre-molar teeth interlocked with the lower to make a highly efficient bite of death.

* * *

The sun moved across the southern sky, and in the mid-morning quietness of the rookery Amik rested, still protected from the wind by the large black rock. She could see nothing because that rock completely blotted out any view of the ocean.

A line of sea urchins marks the narrow strip of land briefly exposed at low tide.

The decline in numbers is evident in the sea birds as well as the seals. The cliffs are still thronged with nesting pairs, but large areas are now empty.

Female fur seals nurse their young on the territory where they themselves were born.

The pups are left for days while their mothers hunt for food many kilometres out in the ocear
Not all survive the time without their mother's milk.

A returning female rides the breakers off shore...

And swims close under the cliffs where murres and kittiwakes are nesting.

Dodging her suitors she climbs onto the rocks, calling to her pup...

...who scrambles over the boulders to meet her.

ne birds of St George: A murre, similar to the European guillemot.

puffin on the cliff-top.

Crested auklets...

...and parakeet auklets.

'ormorants on a high outpost.

laucous gulls forage at the water's edge.

On south-facing cliffs, murres, puffins and auklets sit in the full glare of the low sun.

Throughout the morning skeins of murres stream out to sea, flying in long low lines to their fishing grounds.

She stood up, ears alert, listened again and looked round. All she could see was the quietening activity on the beach and the ten pups close to her, settling down for a restful day. She turned round and began to clamber over one pup and up onto the big rock. She reached the top and looked over. The whack of the howling wind hit her. It was spray and wet and cold, but through it all she could see the sea and white breakers, bouncing, jumping across the cold surface. A wave crashed over the rock.

On land the more experienced fur seals began to move up the beach, away from the buffeting waves. Amik slid back down onto the platform and found shelter again from the biting wind in the lee of the rock. It was blowing up for a big summer storm.

# 5 DEATH IN THE STORM

The wind was howling, it was pouring with rain and the tide was coming in. Amik was trying to snuggle close to the other seal pups. None of them was asleep: they lay fidgeting, waves breaking hard above their heads and spray showering down, all over their sodden fur. The keen Arctic wind howled more viciously, whipping up even larger waves off shore and rocking the hide from side to side. The ashen-faced research assistant was calling up her supervisor on the two-way radio, but he made no response. She wondered why he had not answered. The agreement was that he carried his radio everywhere with him and kept the channel open at all times. She wanted to get back to the home base.

There was a sudden lull in the wind and waves. Elizabeth's teeth chattered. Her hair was sodden, for she had not bothered to cover her head when she had gone out to see what the weather was like—as if she had needed to go out to know that there was a violent storm. Now she was beginning to feel the cold.

The wind howled, waves crashed, pups bleated; all these sounds, intermixed, ebbed and flowed in Amik's ears. A large wave rolled in and broke hard against the large rock on the rookery. Gallons of water tumbled, swirled and flooded down onto the pups. They all started, stood up and moved forward, slithering and slipping over the edge of the platform, and wobbled farther up the shore. The rain beat down on their sodden coats. Some shook themselves like dogs, trying to clear the weight of water from their fur. A third of the animals on the beach were pups. Whole sections of the rookery seemed to be full of writhing black boulders, moving up and away from the sea's edge; the roaring of the wind intermingled with the high-pitched wailings of seal pups waiting for the return of their mothers.

Elizabeth was wondering why she was still in the hide. She could see very little through the teeming rain which clouded the glass of the tiny hut. She could just about make out one bull seal sitting resolutely on his territory surrounded by breakers. The same great waves which crashed over him were almost burying the small pod of squirming pups struggling up the beach. More and more seals were coming ashore, climbing over the rocks, finding shelter away from the furious pounding of the waves.

'How am I going to get back?' she asked herself. The tribike would give her no protection from the wind and rain. It was cold and damp enough in the hide. The wind howled and rocked, picked at some loose pieces of plywood on the roof. If she did not feel safe in her wooden hut, what would it be like outside, riding through the storm on a tribike? She shuddered at the thought.

The eleven pups had kept together. They had moved up the beach to another sheltered rock, out of reach of the waves, and now stood shaking and fidgeting, trying to settle down under the continued flailings of the storm. There was another lull in the wind, and in the quietness Elizabeth heard the sound of a spluttering motor.

'Thank God for that,' she whispered. 'The jeep!'

Down the grassy hill, chugging steadily against the full face of the storm, Alex was butting the jeep, struggling with the steering to keep the vehicle from sliding all over the wet grass. Elizabeth rushed out into the teeth of the gale, half slithered, half fell down the staircase and fought her way across the grass to the jeep. There she pushed herself into the cab and Alex drove away.

'Doctor Alan thought you might need a lift back,' he remarked cheerily. 'We'll collect the tribike after all this has blown away.'

Amik had not eaten for four days. The wind, rain and cold air had reduced the surface temperature of her pelt but she was still warm. The core of her body heat was unaffected, but the energy transfer needed to maintain that heat was draining her personal stores. Death was part of the cycle. Already one pup lay dead, close to the pod, its corpse limp on a low, flat rock. It was not the storm which had killed it; malnutrition and the ravages of hookworm were more to blame. Its mother, delayed by the storm, had not returned on schedule and thus disease and lack of food had made it vulnerable to the appalling weather. Amik carried hookworm, too, deep inside her; the parasites were coming into maturity, sucking her blood and preparing to breed.

* * *

A gunshot sizzled through the sea. One of the men on the trawler was firing. The presence of seals around their boat, eating their fish, damaging their nets, was intensely frustrating to the fishermen and from time to time they tried to rid themselves of the pests. The seals shrieked, whistled terror, and bedlam ensued all around the boat. The female fur seal sped away, swimming fast, twenty-three kilometres an hour. She couldn't maintain that speed for long, but she needed to get away, to outpace the boat in case it should follow her. She had fed well and her internal clock was telling her to swim back to St George. This fright was the spur to return. Ahead of her, although she could not see it from her position low in the water, great swirling storm clouds lay on the horizon, between her and East Reef Rookery. She swam on.

* * *

As soon as the storm subsided, the scavenging fox returned to the beach, and although they had no real need to fear it, all the pups in Amik's pod started up as the creature coughed and padded close by. It came to feed on the small carcass, but the timid pups moved away in a group and hurried back down the rocky beach to the narrow flat platform behind that large, sheltering black rock, close to the sea's edge, leaving the hungry fox to tear at the flesh of the dead pup and return to the oil drum with food. There her cubs were already out of their den and quite capable of eating a little seal flesh.

Early morning, July 20th. Both the scientist and his assistant waited in the hide. The jeep was parked close to the sodden tribike and the catwalk. Dr Alan was anxious. It was time for the return of the first female he had tagged with the TDR. Although he did not really fear for the safety of the female—storms rarely affected mature seals out at sea—he was still worried. She carried his TDR, and the loss of the instrument would be expensive not only in money, but more so in time. It would put his research programme back if the seal or the TDR were lost; he needed the results for processing and analysing. The two of them had come out especially to watch for the return of females after the storm. As always there was a great deal of activity in the early dawning.

They said very little to each other. Elizabeth reported, 'I think that the storm must have gotten to the bull on 34.'

'What? The one right down near East Cliff?'

'Yeh. You can't really see from here.' She pointed. 'Dead. It was the raven told me. Came down for carrion. Don't really know how long it's been there.'

Doctor Alan bit his tongue and thought: For God's sake, doesn't this girl ever keep her eyes open? Then he chided himself for being unreasonable. It was the worry over the return of the TDR, combined with his neck which was playing up again—the old injury. He had got it here on this beach twelve years ago, when he fell trying to mark a bull. God, that had scared him. That bull was one of the ten per cent that don't run away from humans. Those teeth, the foul breath—he still had nightmares about it. Fortunately in those days he had had four assistants. All men. Strong.

'There!' Elizabeth nudged him out of his dream. 'There, in the waves.'

He saw the silver head and a quick flash of red. The TDR strap!

'Damn! Damn! Damn!'

'What?' Elizabeth was confused. She thought he would be pleased to see the female return.

'I wanted to get her before she got to her pup. It's always much more disturbing if you have to take them when they're with their pup.'

'So?'

'Alex's not due till noon.'

'Do you need him?'

'No.' He paused. 'But I'm not too keen on that bull on 15. Really need somebody else down there with me.'

'I'll come.'

'What about your fright the other day?'

She shrugged. 'Experience. I learned a lot, the other day. Especially about myself. I can bang two sticks together as good as the next man.' She looked deep into his eyes, challenged him with the use of the word 'man'. He smiled. His eyes sparkled and he nodded.

'Fine. Come on, then. Quick.'

The adrenalin started to flow. Elizabeth sensed her pulse rate increase. Alan chose two poles from the corner of the hut, one with a noose and one that had a wide-mouthed coarse net at the end. Elizabeth picked the baseball club and the other long pole.

The seal broke surface again and surfed in towards the shore, riding a wave with great confidence. She called briefly, her head held high, before shooting down the face of a large breaker. Amik did not hear her. She was still in her pod although there were now just five of them. The six other pups had gone back to suckle; their mothers had returned. Amik was hungry. She turned on her tail, almost disconsolate, bleated feebly and lay down again. The two scientists hurried down the steps from the hide, the man ahead.

'Stand at the top of the beach. Just keep an eye out. Especially grid 15,' he ordered.

Elizabeth watched him walk slowly down to the edge of the sea. He carefully skirted male territories and dropped to his knees. The drizzle glistened on his green waterproofs. The dark blue bobble hat seemed incongruous, she thought, as she saw him lie down on the wet rocks and begin to crawl slowly towards the female who had rushed out of the waves and paused to orientate herself. The animal looked round and called loudly, waited for a response. Alan was crawling more quickly now. A small wave broke around him. Amik lay still. She had not heard her mother's voice. The mother called again. Alan heard her and jumped up. Throwing aside his hat, he rushed forward, the net gaping open. The female seal turned, wrong-flippered, confused. Faced the sea. She made to escape but Alan was above her, the net quickly over her head, smothering her movements. And now, with the other hand, the noose. He manoeuvred it quickly over the net, dropped the net pole and used both hands to twist the noose, tightening the rope round the neck and shoulders of the female. He pulled her down with the net tight round her jaws. She squirmed, wriggled violently. He jammed the pole into a crack between two rocks and she barked, a sharp, noisy call. Amik heard her, as did the bull on grid 15. Elizabeth caught her breath as the male reared up, rolling a large dark eye.

The scientist's need for the female was, at that moment, greater than her pup's. Somewhere Amik, her pup, was calling. The female heard and endeavoured to reply, but the tightness constricted her and the shrouding net held her down. Elizabeth marvelled at the agility of her supervisor. With one hand he held the seal down with the noose and net, and with the other he pulled out his knife, cut the strap, retrieved the TDR, laid it carefully to one side on a flat rock and began to release the female. Thirty seconds, if that, but enough time for the bull on 15 to stand up and roar. Very carefully, Alan untwisted the noose, whipped off the net and, picking up the TDR, hurried back to Elizabeth.

Breathless, he whispered, 'Come on, up onto the grass, out of harm's way.'

'What about your hat?'

'Tch, forgot about that. Not now. Later, perhaps.'

They retreated to the security of neutral territory, away from all the males.

'There she goes.' Doctor Alan was relieved that the female was making her way back to grid square 14. Amik was calling; the female heard her and replied.

Amik called again, urgently; she had not eaten for five days, but she was not on her mother's territory. The narrow rock platform where she still stood was well away from where her mother chose to be, where she would feed her. Amik had to get back to grid square 14. Not that the yellow lines meant anything to either her or her mother, but she had to get back to her mother's land. She moved forward, small bare flippers flapping a distinctive sound over the harshness of the scoured beach. A female snapped at her, mistaking Amik for an orphan trying to steal milk. Amik ignored the rebuff and edged forward, clambering over brown-pelted females, stained from several days on shore; urged on by the clear calls of her mother, who was herself being harassed by the eager male from grid square 15. The returning silver seal ducked and hurried on. Now the imprinted pup call was drawing them closer together. The bull who occupied all of grid square 14 and part of 13 welcomed her back in his usual way by barking and sniffing at her breath. The female snarled back, barked at the other females who were already there, sharing her land, settled herself again against her rock, and called. Amik heard and bleated, but her path was blocked by the enormous sleeping mound of a male.

It had stopped raining. The wind was keen and fresh, and almost immediately began drying the rocks of the rookery as Amik made a last frantic dash towards her mother. The sun almost broke through the cloud cover and they were nose to nose! Nostril seeking out nostril, breath testing breath, ears picking out each other's close, soft calls. Mother and pup reunited in the unmistakable sound and scent of each other.

The pup shook her head. Her mother sat up tall and waited. Amik felt her way down her mother's neck with her nose, put her left front flipper up against the glistening silver fur. Her mother's thin, furless back flipper moved about the pup's neck, seemed to direct the small head downwards, towards a teat low on the female's abdomen. There Amik nuzzled, sent out a hungry tongue, and began to guzzle. The female seal arched her back, turned almost as if protecting her young pup, murmured a few low sounds, began scratching her left ear with her right back flipper and half closed her eyes.

They were together again.

# 6 ESCAPE FROM THE SUN

Morning glinted at Elizabeth's bedroom window. For once the village of St George was bathed in sunlight; the sea almost blue and the sky bright with white, scudding clouds over a backcloth of glaring azure. Under the grey-green waves off the shore of the East Reef Rookery, the fur seals came looping and rolling, flicking and swishing, lunging and gaping. These were the creatures that Doctor Steinberg had said were solitary, programmed to do just one thing: procreate. He maintained they never did anything other than hunt, swim and bonk. That was what the behaviour research findings proved. Yet out in the breakers, under the water, in and out of the kelp beds, they dived and swam rings round each other, playfully bit out, bumped and nuzzled—juveniles, playing and enjoying themselves in the Arctic summer sea, chasing through icy water and the kelp. Elizabeth watched them, and marvelled.

On the south-facing cliffs, murres in their thousands sat on nests in the full glare of the low sun. Two birds in particular, flapping their wings energetically, were clinging, face in to a fifteen-centimetre ledge, 125 metres above the pounding waves; one edged closer to the other, wing feathers brushing, the breeze from their beating wings dispelling the wreaths of tobacco smoke that drifted across the cliff face. So carefully, so gently, the female eased an elongated, tapered egg from under a pouch of feathers on her lower abdomen, moved aside and the male took over incubation duties. There was no nest, but the egg itself was so shaped that were it to roll it would turn in circles, not fall off the edge into the abyss below. Twenty centimetres to the right of the murres, a parakeet auklet was trying to find cliff space on a three-centimetre ledge. It flapped its short wings in desperation but fell away and was rushed off, out over the sea, by the swirling breeze. Four metres below, but on that same immense outcrop of rock, an Arctic fox walked a well-trodden path. Despite the dizzying height, the creature hurried along a twenty-two centimetre grass-covered ledge, moving with great intent, looking for nests. Suddenly it paused, saw a puffin resting quietly on a narrow pinnacle far out of its burrow. The fox looked down. Kittiwakes, with wide-open wings, soared on air currents, rising in great circles, but all the birds were well out of his reach.

The female murre flew away to hunt the sea, leaving the male firmly in command of the egg. Another wreath of tobacco smoke wafted across, followed by a flurry of stones. They tumbled down the cliff face, banging noisily against bare rock, gaining speed with their descent. A puffin flew off with a deep, growling 'arr'. Another handful of stones and the male murre fidgeted. Lower down the cliff, several kittiwakes were hit by pieces of the gravel and with strident cries of 'kitti-wa-a-k' to add to the general cacophony, left their narrow ledge sites and wheeled round in the sparkling air. Another handful of small pebbles and the murre flew off. Immediately a long pole with a wooden cleft at the end reached down onto that high ledge and deftly picked up the light blue, speckled egg, lifted it through the clear air into the waiting hands of Vladimar, Alex's aged father. The fox looked up, saw the figure of the bespectacled man with a cigarette at the corner of his mouth, and hurried on along his regular pathway. July was the month of plenty. There was food for everybody.

At sea level, close to East Reef Rookery, a jumble of storm-tossed tree trunks lay bleached white, covered in lichen and the dark green stain of moss clumps. Green grass, 15 centimetres high, grew in and out of the rotting limbs and trembled. High-pitched yelps could be heard coming from the centre of the tumbled mass and, stumbling through that grass, jumping over rotting branches, came five sturdy fox cubs. They rolled over each other, jumped up onto the tree trunks, ran, lost their footing, slipped, fell, jumped up again. They were strong and healthy, with soft dark eyes, moist black noses and fluffy coats of intense brown, playing while they waited for food. Their mother was on the beach, pulling hard at the carcass of a newly dead female, who had hauled herself up onto rocks not far from the fox cubs' playground and, exhausted from the trauma of an infected jaw wound which had prevented her from hunting effectively, had lain down and fallen into the deepest sleep of all.

A strong juvenile bull climbed out onto a rock in the breakers. The thick underfur combined with oily secretions from sebaceous glands and support from the guard hairs, all helped prevent the seal's pelt from becoming waterlogged. Now, however, as the young bull shook itself ostentatiously in the bright sun, almost like a dog just emerged from water, it sent a spinning halo of spray around itself, jewels of water sparkling like the fall of a thousand silent diamonds.

Amik was alone again. She had moved back into a pod not too far from grid square 14, but she had also been attracted by the cooling breeze generated by two females fanning themselves. During that clear day of sunshine, the crowded seal colony was awash with the rushing sounds of large, bare front flippers waving, fanning, cooling. Both females were without pups. One was over twenty years old, almost out of the pup-rearing age; the other, fifteen years her junior, was in her first season. Both had mated, both were out of oestrus and both were suffering discomfort from the weather. They sat close together, and for a long moment looked down at the lone pup which had strayed slightly away from the others in the pod and stood, eyes closed, enjoying the sensation of moving air. It was very difficult to find shade from the direct rays of the sun, but Amik let the two females stand between her and the light and concentrated on the deep, dark red glow at the back of her eyes. Overheating was causing obvious discomfort for many of the seals on the beach that day, especially the male on grid square 14. He had not eaten for sixty-three days and lay, immobile, on his favourite rock, both eyes closed. Any movement now was an effort, and sitting all day fanning himself was too tiring.

The old female moved suddenly, barked and snapped at Amik. The alert pup dodged out of the way and ambled back into the full density of the pod, spreading her own front flippers as wide as she could. It helped keep her cool. She was getting bigger, almost eight kilograms in weight and three-quarters of a metre long. Males her age were slightly heavier and longer, but she could look after herself amongst her peers. However, it was now somewhat disturbing in the pod, for there was one strange young pup, newly born and left for the first time by its mother. Amik had snapped harshly at it when it first arrived. It had the scent, the family face, but it was brown, with large patches of yellow, almost white fur in places. A piebald. She had not seen anything like it before. But it snuggled down close to the rest of the pups, seemed to cause no threat. All the same, it unsettled Amik, and from time to time she looked intently at the piebald, watched its movements closely.

Juvenile seals were arriving all the time, surfacing offshore with long necks, stretching their heads high out of the water to get as good a view of the rookery as they possibly could. These were the waters through which her mother had to fight to get to the feeding grounds and then to return to the shore. In the near-freezing breakers, hundreds of juvenile seals, mainly males, rehearsed the movements of the hunt. They enjoyed the constant motion of waves breaking: dived through the surf, waited for the big wave, let it carry them towards the shore. Sometimes the young seals would groom out of the water, but more often than not they would scratch and bite, move air around their pelt, help it filter between the tightly packed hairs to give thermal protection as they swam underwater; but so efficient a water repellent was their coat that their skin never got wet, no matter how hard the scratching or rubbing.

Most of the young males would climb up onto rocks washed by foaming waves, and wait to be challenged. Then they rehearsed the rituals of territorial possession, barking, snapping, lunging. The older the male, the more difficult it was to dislodge him. Even to the pup, watching the teeming activity at the edge of the sea from a distance of ten metres, it was obvious that it was the brownish-black creatures, who had developed a shaggy mane around their shoulders, who held reign for longest, not the younger, slimmer, silver seals.

Like the females, most of these young males had regular feeding cycles. Not for them the fasting of male adulthood. Only twenty per cent of the juvenile male population was on shore at any one time; the rest were either feeding at the edge of the continental shelf, two hundred kilometres away, or swimming offshore. However, there were a few of the older juveniles who spent more and more time sitting on rocks at the shoreline, waiting, watching, staring intently at the activity on the rookery itself.

The big male on grid square 14, looking much thinner now, stood up and shook his dry, shaggy mane. A large, beady brown eye rolled round. He looked over his shoulder. It was a walk of no more than ten metres to the water. Most of the other seals on the beach were lying down, fanning themselves. Pups bleated, a distant raven called, and from the direction of East Cliff came the echoing, haunting cry of a rock sandpiper. Slowly at first, but with increasing speed, the male vacated his territory and slid out of the heat of the day into the cold waters of the Bering Sea. Without pausing, he plunged headlong into the breakers, submerged, pushed hard, straightened up underwater, flicked a right flipper, curved in a graceful arch, opened his mouth, let go a flurry of bubbles, pushed again and shot to the surface; there he blew loud and wide, shook his head, made a heavy, slow porpoise that hardly lifted him above the surface, and swam slowly away from St George Island.

One young male, on a rock in the waves, saw the territorial bull go from the beach and cautiously edged out of the sea and up the rocks of the rookery, crossed the yellow-painted line, slithered over a rock with 14 painted clearly on its side, and approached a group of females. They snapped, barked, waved heads on sinuous necks, and he kept his distance. Settling down on a hot rock, he began fanning himself. The heat seemed intense to him. It was seventeen degrees Centigrade.

Elizabeth had taken her parka off, it was so warm sitting there in the hide. In her notebook she marked off the retreat of the male. Doctor Steinberg was sitting at her side. He had retrieved four TDRs and had the computer printout in his hand. He was quite proud of the findings and was explaining them to her.

'The ocean's a continuation of the earth's atmosphere, with its own patterns—winds and breezes, areas of warmth and coolness, high and low pressures—and it sustains its own unique life forms, which can be divided conveniently into three types, plankton, nekton and benthos.'

Elizabeth knew all this. After all, she might be an undergraduate research assistant earning a few extra bucks during her summer vacation, but she was majoring in oceanography.

'The plankton,' he continued, 'include those plants (phyto-plankton) and animals (zooplankton) which either float passively or have such limited powers that they are carried from place to place by the currents. The nekton's all the marine animals like squid, cuttlefish, seals and whales. And the benthic population inhabits the ocean floor—you know, worms, seaweeds, molluscs, sponges, sea anemones, that kind of thing.' He rested the computer paper on the narrow ledge and leaned across her to pick up a TDR. 'Now this little contraption measures water temperature. It's the first one that's ever done that.' He smiled. Doctor Alan very rarely smiled, she thought.

'Anyway,' he continued, 'the boundary between the sunlit surface waters of the sea and the twilight areas . . .'

'The dysphotic zone,' she interrupted.

'Yes,' he said, somewhat surprised.

She wanted to hint to him that she knew a lot more than he was giving her credit for. Not that he seemed able to take the hint. Actually, for a pure research scientist he was a very good teacher.

'So then, as you probably know, the dysphotic zone is located somewhere between 30 and 60 metres. The variation in depth depends upon latitude and the density of living organisms suspended in the water. Many of the defenceless creatures hover around the boundary of the dysphotic zone during the day.' He paused, changed the tone of his voice. His little joke, she thought. 'It's safer to be in the dark when predators are around. But when the sun goes down there's a movement upwards towards the surface. That's why fur seals tend to eat at night.'

He had attached the new TDRs to his portable computer, and now the printout gave him evidence of water temperature for the first time. It seemed to support his theory that the seals fed where cold and warm water mixed, around the edge of the continental shelf.

'The problem is, Elizabeth . . .' His hand went to the stubble on his chin and he played slowly with his fingers across the rough surface. 'This decline, I know we think it's halted now. But for how long? I mean, up to the present century the cycle of life in the ocean was very efficient, delicately balanced but efficient. Marine organisms died, sank to the bottom, decomposed and returned all the elements which they'd consumed during their lifetime, back to the ocean. But these trawlers and factory ships, what are they doing?'

He never really looked her in the eyes. Now his keen blue eyes were gazing out through the window of the hide, over the rookery, out deep into the distant haze of the Bering Sea. That was where the answer lay, in the open ocean. She knew that, but he was still talking and she had to concentrate.

'As a rule nitrates and phosphates accumulate in winter and then are used up rapidly in the spring. Without them plant life disappears. Tch! What am I doing here? Giving you a lecture, eh!'

'Sure, but it helps me understand.'

'Does it?' He seemed to need that reassurance.

'It's nice to know what I'm doing.' A feeble thing to say, she thought, but it was all she could think of.

The grassy bank at the top of the beach was warm and the lemmings were nibbling nervously at sedge, their eyes half closed against the bright rays of the unclouded sun. The fox was downwind, tail curved but pointing sharply down against the ground. Imperceptibly the animal's back moved slowly into an arch and it raised itself onto its front pads. It held itself for a moment in that almost awkward position, then made a high, final leap onto its prey. The remaining lemmings shot back into the nest in between the cracks in the black boulders. The grassy bank seemed deserted as the fox trotted away, the dead lemming in its jaws.

# 7 TO THE EDGE OF THE SEA

Off the East Reef Rookery a stellar sea lion was patrolling. At least it was swimming up and down, at a leisurely pace, occasionally breaking surface, rolling over, pausing and looking inshore. It was fast approaching the time when it would be able to feed at leisure off fur seal pups caught unawares in the kelp beds. Elizabeth, however, had no eyes for that creature; she was being kept very busy watching out for yearlings. Juvenile males were now coming ashore onto the hauling grounds in scores, and as Alan had marked a fair few pups last year, putting small numbered plastic tags on their right front flipper close to where it joined the abdomen, he wanted Elizabeth to check all returning juveniles, to see how many yearlings returned and when. Late July was very early for a yearling to return to the breeding colony—Elizabeth knew that—but it was important to know. Most pups during their first winter swam a fair distance south; not as far as mature females, but far enough to feed well, and then slowly they returned. Some took a lot longer than others, not arriving back until October or November, and then they hardly stayed onshore at St George. Those yearling that arrived early, however, were likely to be the strong ones. They would gain more experience of life on the rookery, very important for a male. Any observations Elizabeth might make could throw light on changes in breeding success or variations in feeding patterns. So, in addition to charting the return of females to the marked squares and the birth of pups, the young researcher was looking out for marked juveniles as they pulled themselves up onto the hauling grounds to the east and behind the fur seal breeding colony.

Elizabeth took a short break from her observations. Standing up, she stretched, and opened the door of the hide. In the relative quietness at the back of the beach, away from the noise of the rookery and the cacophony of the sea bird colonies on the cliffs, she picked out the brief, fluting 'turee-turee-turee-turiwee' song of a snow bunting, and realised that the chicks, which only a few days ago were still cheeping from lowly perches on debris around the base of the hide, begging food from their mother, were no longer anywhere to be seen. Flown, she assumed, building up their strength for migration, in September, to their winter feeding grounds. Where they found all the insects to feed off Elizabeth did not know. There seemed so few on St George and, fortunately, no mosquitoes. That was such a relief. Before she had come to the Pribilofs she had thought of the tundra and all those dreadful mosquitoes. On St George, she had not been bitten once! She closed the door quietly and returned to her seat. Focusing her binoculars, she searched for a small fur seal pup. Not Amik—she had been overlooked by Elizabeth as a source of interest, once she had been recorded as a birth. It was the piebald pup that Elizabeth had fallen for and spent a great deal of time talking about to Doctor Alan and Alex.

'Bleach Boy—' that was the name she had coined for the piebald. 'His mum's just come back, started feeding him again.' She found that she had begun to talk to herself. Meanwhile Amik's mother was still out at sea. She was just skirting a school of dolphin, while down on the beach Amik was in a pod of six, hungry but strong, almost glowing in the warm light of evening. She fanned her flipper and bleated. The pangs of hunger were biting deep into her belly, and it was a big one. During the nursing period a pup's stomach occupies most of the body cavity. Amik looked up as the raven fluttered in and landed not far from her.

Elizabeth squinted down the telescope, and focused up, full on the face of Amik. Sucking lice, how foul! My God, it was disgusting. She did not really think that, because she was a hardened natural history observer, used to all sorts of sights, but deep down she always thought it. There in her lens she could see them, on the eyelid and round Amik's eyes. She knew about parasites, thought back to one of Doctor Alan's explanations:

'*Antaretophthirius callorhini* were found on sixty-seven out of seventy-five seals examined. All the pups examined were infested, and there was another type of lice in the fur-covered areas, *Proechinophthirus fluctus*. The interesting thing was that the number of lice on any seal decreased with age. The pups seem to have the roughest deal, get all the parasites off their mothers.'

Deep down below the waves, fifty kilometres out in the Bering Sea, there was the sound of the mournful chiming of a sunken bell, slightly cracked. The female seal, returning from her latest hunting trip, heard it. She sped on. Shrimps were clattering all around her. As the bell faded into the distance there was a new sound, of wood-pecking, then the light tapping of a timpani—in fact the variety, the sounds of almost the whole percussion section of a symphony orchestra. The seal heard the noises and recognised their source—she was sharing the waters with walruses!

It was late in the afternoon. The seal would make landfall at dusk, would begin to feed Amik as the sun dipped below the horizon. Her long, streamlined mammary glands were full, almost aching to be milked again by the hungry lapping tongue of her pup. The sun sparkled and the sea shone bright, clear into her eyes; glinting, causing her to squint. She did not need to look to know where she swam. She dipped below the surface, bubbles racing past her alert ears. It was a direct line back to St George. The colour of her fur was not so pure now. All winter, while at sea, after her moult she was truly silver grey, but the days on shore reflected in her pelt: six days' hunting in the open sea had not removed all the yellow-brown staining which she had picked up since the birth of her pup by lying in and moving through the mud and excrement of the crowded rookery. She swam on, broke surface, and with long, deliberate strokes from her powerful flippers, propelled by strong muscles in her neck and shoulders, she moved forward at a steady speed of fourteen and a half kilometres an hour.

Above her head, clouds began to form. It looked as if she was journeying through great canyons in the sky. To her left, about two hundred metres away, a dolphin jumped, followed by a second and a third. They had come together to hunt and feed. Now, towards the end of the afternoon, they socialised, swimming together in small, ever-changing sub-groups. Touching, caressing each other with their flippers, swimming belly to belly and poking noses at each other's sides. She swam fast around them, circling the school, for although they were not predators of fur seals, she disliked their playful curiosity and she urgently needed to return to St George.

Evening was returning to the beach. Amik scratched noisily at her fur. More and more females were landing. The salt smell, the noise of the breakers, the constant movement of seals in and out of the water, beckoned the pup. She had explored most of the rocks within a thirty-metre radius of grid square 14, but she had never been down, not right down into the waves. The tide was out. She shuffled down towards the quiet, lapping waves. A large log pounded dully against the shore, the waterlogged timber sounding like a muffled bell against the hardness of the rocks. Occasionally there was also a crunching sound as the tree trunk crushed a barnacle. Spume, whipped up by wave action against the irregular shoreline, was floating white, catching at the edge of the black rocks, and a large piece of discarded blue net clung to some small, spiky, brittle spheres. There was a mat of these things right at the edge of the waves where the tide turned weakly from its ebbing. They were about ten centimetres across, pale green with violet tips to their strong, short spines, and in addition to the blue plastic netting they were covered in bits of seaweed which linked the creatures together so that they formed almost a carpet, lining the edge of the sea. The pup knew about barnacles and limpets, her tongue and her nose had explored them, even the jelly-like blobs of sea anenomes; but these creatures were new to her.

A heavy, fairly mature juvenile male floundered out of the waves, crunched across the mat of sea urchins and up onto a rock. The sound intrigued Amik; she looked again at the spheres. The digits on her front flipper caught at the blue net as she made to push one. She pulled back, safe from the ensnaring lure of the blue plastic. Not too far away her mother was calling. Immediately there were no other thoughts or distractions; she turned and clambered as quickly as her small flippers would take her, back to the big rock on grid square 14.

* * *

Summer came speedily. Flowers bloomed, the halibut boats were out in the grey waters, the island children returned from school in Anchorage and the tourists arrived. Every year Alex showed up to twenty-four tourists around the island in groups of six. They came for four days at a time to see the seals and take photographs of the birds. To help Alex out, Dr Steinberg always gave them a lecture. Elizabeth had gone along to the small island hotel where the tourists were accommodated and listened to her supervisor's talk.

He was summing up. Whether his talk was appreciated Elizabeth could not tell, but at least they were getting the most up-to-date information available about the fur seals of St George.

'These animals spend eighty to ninety per cent of their lives at sea, and because the reasons causing the decline appear to be out there in the open ocean, we must investigate their life out there. The trouble is that research from a ship is expensive, so we've got to use land-controlled instruments. I've already used TDR. That shows us the number, duration and time of day when diving occurs. Long-term comparison will show whether the seals have to work harder to feed now than in the past. Then I've started to measure swim velocity, to show the foraging range of the seals and the means by which they capture their prey—whether it is burst swimming or sustained speed. Another new instrument is the CTD—conductivity, temperature and depth profiles on each dive. This is exciting research.'

Unconsciously, Elizabeth yawned. It was not intentional or indicative of boredom but it caught Alan's eye. He said nothing, but continued. 'Yes,' he said, trying to inject enthusiasm into his voice, 'it's exciting because for the first time it lets us study the relationship between a marine mammal and the oceanographic processes that structure the sea. What I'd really like to do is make use of the new satellite transmitters so we can track the animals through the high seas. That would give us much more information on their entire migratory route.'

'Thanks, Dr Steinberg,' said Alex. 'Any questions?'

On the beach there was a great deal of activity. Ruddy turnstones pecked and flew, flew and pecked in a cycle of fidgeting movements down the line of weakly breaking waves. Glaucous gulls chased each other across the surface of the sea, trying to capture some of the offal thrown overboard by a halibut boat that chugged noisily past the rookery. And the raven glided down, alighting with a noisy flutter of its wide black wings, close to a big black rock where a growing fur seal pup was hungrily sucking milk from a newly returned female.

# 8 SEA LIONS!

In the high Arctic, hundreds of miles to the north of St George, the sea ice had begun to reform and creep south. There was an eerie quietness about the island. The cliff ledges, the puffin burrows, even the high rocky face where the least auklets roosted, all were empty. The birds had flown south. An unusual autumn twilight was falling, cold and ominous, glinting waters in the setting sun. The lightly clouded sky held a flush of fireglow, florid coral-pink blushes, almost the colour of lobster or salmon. The mature male fur seals had already left the Pribilofs for their winter hunting grounds in the Bering Sea, and in the bay beyond the East Cliff Rookery, stellar sea lions prepared.

On the rookery, in nooks and crannies between the rocks, were vestiges of fur—seal fur. Some hairs blew along the beach in the wind, some floated at the surface of the sea, eddying round and round in quiet pools at the tideline itself. The summer moult had taken place. Most of the seals had renewed their coats for the coming winter. Amik was now a silver seal herself; she had lost her black puppy coat and was almost indistinguishable from her mother. She was big, too: eighteen kilograms—half the weight of her mother but only fifty centimetres shorter.

The darkness, which gathered in the lee of High Bluff and oozed down onto the village of St George, clung close to an old man sitting on the small cannon which stood on the grass close to the church. He pulled at a cheap Russian cigarette which his son had brought back from Dutch Harbour. Smoking was not good for him, Alex told him that. Yet it was Alex who had brought him the cigarettes. Such were the contradictions in life.

He was an Unungan—Aleut, the Americans called them. It had been necessary for the fur hunters to enslave the Unungan people, bring them north from their homeland on the Aleutian Chain and make them labourers in the harvest of the seals. The Russians had taken the Unungan religion away from them, the religion which relied on the sun and the moon and every living thing with a purpose. The fur hunters had given them one God and a church. There it stood, a structure of white weather-boarding overlooking the village. One of the first things the city council had done when the Americans gave them the twenty million-dollar settlement was put a new copper cupola on the church. It had taken them twenty-seven years of lawyers and courts to prove that Unungans on the Pribilofs had been mistreated by the United States, and when they had proved that, then animal protectionists came along and said that they, the Unungans, had no culture and that the killing of fur seals must stop. But it was the quality of Unungan culture that kept his people going through all the hard times. Patience, perseverence, justice and an indomitable will to survive in the natural environment. That was their culture.

His great-grandfather had lived in a sod-walled hut, partly dug in the ground, dark and dirty because of the soot left from the burning of seal blubber. That was in the time of the Russians. When the Americans came his great-grandfather had petitioned the new fur seal company to let him build his own home, on land reserved for him as an Unungan. But the Americans thought that was devious, a scheme to establish the Unungans' permanent right to live on the Pribilofs. They had said no. Eventually, when he was a child, they gave the Unungans the right, let them build the weatherboarded cottages. Now the cottages were fifty years old and needed attention. And the government buildings, which would soon be their buildings, were also in disrepair because of yearly government funding cuts over the last ten years, and most of the public equipment too, even the airstrip, needed overhauling. And there were no funds, not even to build new homes for the young Unungans who wanted to stay on the Pribilofs.

His left hand was cramped. He drew it up tight to his face, looked at it in disgust.

'What kind of a hand is that?' he asked aloud. 'A claw?' For a moment there flashed into his mind the overwhelming smell of salt and seal blubber, the skin plant and the pain of his hand, seeing it trapped in that roller. That was thirty years ago when they killed thousands of seals every year.

He finished his cigarette, threw the stub away and lit another.

Now the Russians did not have a God. They taught the Unungan children that at school in Anchorage, such a long way away. Taking their children all that way, away from their homes for most of the year, so that they could school them in American ways. The Americans had a God, but Alex said it was not the God of the Unungans' church, the Russian Orthodox Church. When his great-great-grandfather came to St George, the Unungans had had to make their next new culture very quickly. Living on this misty island, killing fur seals, skinning them, they had made up a complete ritual dependent on the arrival of the fur seals in April and the killing of the young males in summer. That had all stopped. Unungan destiny had been inextricably linked with the fate of the fur seal since the first Russian set foot on Unimak and made his people slaves. Now the tie to the seals had been broken, by the Government. The killing of the fur seals had stopped.

He screwed up his eyes until they were just slits. He was thin and gaunt, with deep wrinkles in the back of his neck, and his skin was sallow. Not white from lack of exposure to the elements, but untanned from the sun because he rarely saw it peep out from behind the mist and clouds and he always wore thick clothes to keep him warm in the cold northern air.

Yes, across the dark waters he could see the small shadowed heads swimming away, back south; leaving the shores of his island empty again as they always were in winter. The long dark nights were returning, full of ice cracking in the frozen sea and the wind blowing over the deserted cliffs.

Alex walked up from the post office, blowing hard. That boy rushed around too much, he thought.

'Just been phoning through to Seattle,' Alex informed him. 'Doctor Alan wanted me to order a few things for him.'

'He'll be off soon.' It was half a question.

'Yes,' replied Alex.

'And that girl.' Vladimar never seemed to ask questions; always made statements which could be corrected.

'Liz?'

'Yes.'

'Yes, she'll go off with the doctor.'

Alex looked down at his father, sitting on the cannon. It was a tired conversation, both struggling to find words while their thoughts were elsewhere. Both of them, father and son, looked far out to sea. Neither knew what the other was thinking and yet both sensed it was about the fur seal migration.

'There's not enough fish,'

'What?'

'Out there, for the seals. On their way back.'

'Why not?'

'These trawlers.'

'Oh.' And Alex remembered. Five, six years ago, he had gone out with the Taiwanese boats to learn about the factory ships, the big ships. He had been trained for the deep sea fishing.

'They didn't tell you what happens when you take the fish away, leave the sea empty.'

'No. But the boats don't leave the sea empty.'

'Fish don't grow again. It takes time.'

Alex looked at his father. Read his mind. He was right. Now the big boats came right up, close to the Pribilofs. And they, the Unungans, were building a harbour for them; for the foreign fishermen from the other side of the world, so that other people could come and take the very fish, their fish, out of the sea, their sea. The foreigners were island people, too. But Japan and Taiwan. Why did not the trawlermen fish there, in their own ocean?

'No one. No one knows,' Alex's father murmured. 'We used to. Us. The Unungans. We knew. But they don't listen to us. I remember 1956.'

Alex knew the story well. His father told it often. He had been only a child. It was the year the Americans ordered the killing of the sacred cow, the female fur seals.

'I remember the beaches covered, littered with the bodies of starved pups. You never kill a female, that's what we were always taught. Nasty business.'

His father was right. As Alex had grown up, every year it was the same. Females killed and pups dead from starvation on the beach.

'They stopped the killing, but the seals never recovered. Now it's the fish they take.'

'But, Dad, it's not that much, not round here.'

'They said, kill the females and the seal herd will grow.'

Doctor Alan had explained it once to Alex. He had said that the decision was based on the belief that the herd was out of balance, too many females for its own good. His father was right. The increase in productivity which should have happened as a result of the policy of culling females never took place.

'But that's got nothing to do with the seal decline. They stopped killing females in 1968. Dr Alan said that the decline carried on a long time after they'd stopped the killing.'

'You can't believe any of them. Taking the fish out of the seas is like taking it out of the mouths of the seals. And us. We don't need the seals any more, but the fish for ourselves.'

All the Pribilof islanders knew about the fur seals, that there was something wrong. The scientists might find things out by sitting around and watching them all day, but the Unungans *knew*. Old men often told of the times when the rookeries were teeming. Even Alex remembered mature adult males climbing up, right out of the sea, from the north rookery down by the harbour, not far from the skin plant, and waddling up the main street of St George. Once there had been a dance and somehow one of them had got right into the old community hall. How they'd laughed. And he and Greg and Agafon had chased it back down to the sea.

'But pups are the same weight now in September as they always have been,' he heard himself saying. 'The doctor says so. And the females don't spend any extra time hunting. I know that. You know that. All the doctor's figures prove that.'

Alex had to argue this way. He supported the harbour project. The island needed the fishing to survive. But deep down he was afraid. Programmes he had seen on TV and magazines he had read—it was balances. If there was less food available creatures produced fewer young. Only recently, there was this programme about owls and the number of voles. If one year there were more voles available, then there were more baby owls; if there were fewer voles because a bad winter had killed them off, then there were fewer owl chicks. It was simple. He shrugged the thought off, tapped his father gently on his leg.

'Come on,' he said. 'Sitting here in the damp is no good for your chest. And don't go lighting another cigarette, either.'

* * *

On the East Reef Rookery, as always with dusk, activity level increased. A white Arctic fox stalked the sea's edge, clear against the blackness of the rocks. Unusual, that fox, because over ninety per cent of St George foxes stayed blue all the year round. It sniffed and searched, but found nothing. All the fur seal pups which had survived were now big, and none of them waited in patient, anxious pods while their mothers were out at sea hunting. Now groups of young seals made for the sea. There, Amik found a grace and freedom she had never had before, and in the sea she practised the most essential skill of all—hunting.

Above her head when she dived below the waves was the light, the surface reflecting the sky, shadows rippling. Amik learned to see, to streak fast and hard through the water, attack the surface, twist, turn, thresh and lunge. She came out into an open space of light-filled water—a break in the kelp canopy. In the centre of that space was a small sharp shape, black against the surface of the sea above her head. Amik stalled her flight through the water and held herself steady. Then she pushed powerfully with her front flippers and raced hard ahead and upwards, direct at that black shadow. It bobbed gently in the waves. There was no escaping. Amik took it full in her mouth. Caught it! Her first catch. Seaweed! It might have been a fish.

There were many rehearsals that autumn—hundred of pups, all in the water, swimming from instinct, hunting from inbuilt memory dragged out of the æons of hunting experience of previous generations. Sometimes Amik chased weak sunlight on the water and caught it, her teeth biting deep into the invisible. Once she chased and caught a dead fish, felt the cold scales, the bones. She carried it to the surface, tossed her head and threw the fish, porpoised after it, caught it again. Her sharp teeth tore the flesh, giving food for the chasing glaucous gulls. Amik was no match for their harryings. As soon as the large birds recognised her find, they swooped over her head and she dived, sank gently back into the kelp, leaving them to wail their quarrels and snaffle her find.

However, for all her hunting while she stayed on St George, Amik never ate flesh; it was her mother's milk on which she still depended. Or was it? The female had left her nine days before and she, Amik, had spent most of those days in and out of the sea, swimming, weaving, chasing, surfing, floating and porpoising, often sharing these activities with the piebald. He too had moulted, but not into a silver seal. He still carried a piebald effect, not so pronounced as in his first coat, for where there had been a dark brown and off-yellow striped effect, now it was much lighter; but nonetheless he was piebald and very different from the true colours of a silver seal.

All the pups swam instinctively. There had been no teaching, no learning; as soon as the first wave washed over a pup's head it knew what to do, and the more it did it, the more confident and adept were its movements in the water. Viewed at a distance and from above, the pups looked like enormous tadpoles threshing about in crowds at the edge of a gigantic pond, the ocean. There were thousands of them in the water, and females returning to land, from hunting trips at the distant feeding grounds, no longer had to fight off the advances of love-hungry males; instead it was the turbulent confusion of seal pups preparing to migrate. And once on land the mothers had to call and call, had to wait for their adventurous pups to hear them, swim to shore and find them; because despite their rehearsals of hunting techniques and feeding ploys, all the pups still relied on their mothers' milk. In all his years studying fur seals at St George, Doctor Alan had never seen a pup eat anything solid apart from gravel.

Half a mile away to the east of the cavorting pups, a group of stellar sea lions put to sea. Powerful strokes of their front flippers carried their reddish-brown heads through the darkening waters. The heavy manes about the necks of the males created a dull wake across the calm surface of the sea. With purpose they rounded the headland, swam straight across the small bay, ignoring East Cliff Rookery, and headed direct for the teeming waters off East Reef. There, about one hundred and fifty metres offshore, due north of the eastern hauling ground, eight heavy heads submerged.

The fur seal pups were floundering about in mock fights, in and out of a dense mat of floating kelp. Shrimps clapped and clattered noisily, waves rushed bubbles through the seaweed; distant, hardly heard in the background, was the wail of movement of a hunting sea lion diving, swimming efficiently underwater. Elizabeth had decided the light was fading too fast and observations were becoming impossible, so she was packing up, pulling on her gloves and getting ready to leave. She shut the window of the hide and, on account of the gloom, failed to see the confused threshing of surface waters, thrashing flippers trying to flee the turmoil.

Through the kelp they had come, into the sea not far away from grid square 14—eight marauders, come to feed. Stellar sea lions are close relatives of the northern fur seal, and one of the most common predators of northern fur seal pups. There was a confusion of tight, strong neck muscles, wide gaping jaws, sharp cutting teeth; of lunging and roaring, blood dripping from wounded sides, pups fleeing in all directions. A whack of water hit Amik sideways. A reddish brown flank turned with all the force of adulthood, a back flipper caught the pup full in the face, blinding her for a moment. There was a roar of movement to her right side and a sudden pain at her flank. A flooding blood-red stain filled the water cavities between the kelp, and the surface of the ocean was a banging, violent furore. One male stellar was shaking a fur seal corpse from side to side, the lifeless pup held tight in its canine teeth.

But it was not Amik, it was her companion, the light-coloured male juvenile, easier to see in the kelp, easier to focus on as the target for the last, fatal lunge of a predator. And while the sea lion fed on the piebald, Amik was porpoising, swimming as fast as her front flippers would carry her, heading south; and close to her, another four pups, all with the same intent, fleeing the dangerous waters of St George, heading towards fish-rich grounds and safety. Without pausing they swam on, slowing their strokes but swimming steadily through the dark, empty ocean. Only once did Amik pause to lick her wound.

The sea lions turned to make another foray through the waters still crowded with pups, but Amik was away, out at sea, in the deep waters away from the island, swimming free.

Suddenly there was a sound—eerie, echoing, a sound she had

never heard before. And another, that of a submarine sonar penetrating the heavy waters and then coming closer. Again it echoed, underwater and approaching the pups, or were the pups approaching the noise? Again and again, and then a full boost of other sounds—moos, chirps, whistles, clangs, the general din of a barnyard full of scores of animals all competing for attention. The pups were disorientated and slowed down. In front of them, right across their swimpath, were ghostly white shapes, sixteen feet long. More din, a whistling like a caged bird and the occasional startled sound of a woken piglet and a squeaking five-barred gate. Beluga whales. They passed by. Amik watched them go, observed them closely, knowing them by acquaintance and storing up the knowledge: of the white creature with its pronounced forehead melon and its unique facial expression which, head-on and close to, on account of its wide, flexible mouth, looked for all the world like a curious smile.

For three full days the small group of seal pups swam south. The salt healed Amik's wound; the rich waters, full of krill and small shoaling fish, yielded up easy food; and the cold, rushing currents pushed them onwards, streaming in migration with thousands of other pups and mature females, all separate, all independent, but all moving in the same direction: south.

All this time the female seal, Amik's mother, had been hunting at the edge of the continental shelf; now she returned to St George and called as she always called, lifting her head out of the breaking waves, surfing in onto the beach, calling again, and moving hurriedly to the black rock on grid square 14, calling occasionally as she crossed the rookery. There were five other females close to her land but there was no territorial male. She settled down on her usual spot and called again. Amik was slow in coming. She called once more. Recently, certainly on her last visit to shore, Amik had been a long time returning, and when the pup had finally arrived she was wet from the sea. Amik was growing up.

The female lay down, closed her eyes and called again. She waited. There was no red glow in her mind. Night had already fallen, black and complete; thick cloud hid the moon and stars. She waited for several minutes and called again. It was a long night for Amik's mother. She dozed as she usually did on returning, and from time to time she called. Her teats ached to be sucked but her milk oozed unused onto the rock. There was no Amik. In the morning, as soon as it was light, she stood and called a long, clear call. Then, rearing up to her full height, she climbed up onto the big black rock and called again, looking intently up and down the beach. There were pups, but not hers. And so it was for the whole day. She called intermittently but with no response, and rested meanwhile on the hard rock, more milk oozing away.

Eventually, as the sun sank below the horizon, she moved slowly down to the edge of the sea. She was still full of milk and her walk over the last few rocks down to the breakers seemed very laborious, almost weary. Through a gap between two big rocks there was a surge of a seventh wave, a big one, cold and clear. It washed over her. She was submerged. Without a look back, immediately she slid forward and dived. She was back in her element. Without any apparent effort, but with the noisy stridulations of crabs in her ears and the incessant chattering of shrimps about her head, she swam forward into the floating forest of wavering kelp and disappeared into the grey, murky depths of the Bering Sea. She never saw her daughter again.

* * *

Amik swam on. Southwards. The hunting was not easy and the sounds, the noises of the sea, were frighteningly new, requiring time-consuming investigation. There was nowhere to hide, apart from the floating kelp, no rocks to climb back onto, no warm teat to suckle and no sheltering female form to hide behind. Amik was on her own. Her wound was fast becoming a scar. She had survived her first real test; now her future depended entirely on her skill and her own efforts.

Although she did not know it, there were fewer fish in the ocean as she and her four travelling companions swam slowly through the southern reaches of the Bering Sea. Each day the seal pups heard the distant but unrecognised sound of a trawler's engine heading out of Dutch Harbour, on Unalaska Island, north and west to the fishing grounds at the edge of the continental shelf between Siberia and Alaska. Instinct was driving Amik onwards towards Unimak Passage, into the roaring waters of the straits to the east of Unalaska Island, on into the wide waters of the open Pacific.

It was juvenile seals who were not surviving. According to Doctor Alan and his colleagues, Amik and pups like her were most at risk, especially on their first journey south. It was with these creatures, the young, immature fur seals, that most scientists felt the key to the decline lay. Amik was swimming on along the narrow corridor of migration, following swimways used by her kind for millions of years; fish-rich corridors that man had happened on in the last twenty years and had begun to empty, scooping fish from the water with greater and greater efficiency and with little seeming thought of the effects on tomorrow.

Back on the island Alan and Elizabeth spent two days packing and tidying up. The young undergraduate was increasingly frustrated with her supervisor's inability to provide her with a succinct and final answer to the cause of the decline of the fur seal herd. To her impatient mind it seemed simple. She wanted answers.

'Surely you know what caused the decline?'

'Honestly, Liz, I don't.'

She sighed and continued to fold up the yards of computer printout with which Alan had been working.

'All these statistics. Information from the TDRs.'

'They're only the beginning. Look—' there was an element of impatience in Alan's voice. 'You answer this question: How can the condition of the average seal, in terms of growth rate and size, be improving while the seal population is declining?'

'I thought you said it wasn't declining.'

'Well, you know what I mean.'

'All right, I will answer it.' She put the paper down and smiled with determination. 'First you kill three hundred thousand young, mature female seals over twelve years. Then you fish the seas until they're nearly empty. Only the fittest, healthiest seals can survive, but in decreasing numbers. It's a question of balance. All wildlife responds to abundance or scarcity. Look at the lemmings.'

'That's just assumptions, jumped-at conclusions. You've no evidence for any of that, Liz.'

'It's bloody obvious to me,' murmured Elizabeth.

'he scavengers of St George – Arctic foxes in their summer coats – prowling among the rocks or afterbirths and dead seals...

' scaling the cliffs in search of birds' eggs.

In the short sub-Arctic summer, the grassy cliffs are bright with flowers blooming bravely in
cold, damp air.

ɴn the rare sunny days, the unaccustomed heat can be too much for cold-blooded animals.
ɬippers make excellent fans.

ɡh Bluff, seen from Seal Lake.

The hide – more like a garden shed on stilts – where every summer the scientists watch an monitor the activity in East Reef Rookery.

The skin plant, echoing with memories of the days when 200 Aleuts worked to prepare the se skins for the world markets – now a garage for tribikes and jeeps.

Doctor Alan appeared not to have heard her last comment. He said, 'It's difficult to link cause and effect when you've got far-ranging marine mammals with two sexes, and about twenty-five age classes. OK, it's environmental, whatever that means. And it's survival rates, not fertility, and juveniles rather than adults or pups on the beach, and it's in the open ocean.'

'Something's getting them out there.' Elizabeth tried to inject a jokey irony into her voice.

'And it's more than sea lions.' Alan had half a chuckle, too.

But Liz was concerned. Surely, after all these years of research, scientists should have come up with a reason.

'We haven't enough hard data,' continued Alan. 'I know what you're thinking: by now there should be a solution. Well, there isn't. In the last twenty years or so there have been changes in the climate in the ocean. The surface of the North Pacific seems to have got colder. Perhaps that's where the juveniles find it difficult, because, especially in their first winter, they have to learn to survive by trial and error. We just don't know anything about the feeding habits of juveniles.'

'And then there's all this gill net fishing.'

'That too. It all contributes. But if you want a tidy answer there isn't one.'

'Will there ever be?'

Alan shrugged his shoulders. Elizabeth went back to her packing, thinking to herself that as long as men interfered in the ocean, upset the balances, the animals would always be threatened. Alan was all right, but it was all so political. Who was going to stop the Japanese and the Taiwanese over-fishing? Alan's department would not have the courage to point the finger; anyway, scientists were perfectionists and there would never be one answer to the cause of the decline. And in her most cynical moments she thought, he wants money for his satellite tracking, so he's not going to say he thinks he knows. Not just yet.

They went out to East Reef Rookery for the last time, to get the hide ready for winter. They had taken out the glass windows and were boarding up the openings and screwing down the door.

'I don't expect there'll be much left after the winter,' he said. 'But I always hope.'

'Has it ever been here when you got back?' she asked.

'Yes,' he smiled ironically, 'but always in pieces.'

There were still a few seals on the beach. 'Mainly yearlings,' Alan explained. 'Last year's pups, just got back. They'll spend a few days here then turn round and swim south again.'

She had come on the tribike and the doctor in the jeep, so to Elizabeth it was a short, sad procession that Doctor Alan's jeep led back to the field station. They flew out the following day, back to Seattle—Alan to his laboratory and the continuing fight for more funds to pay for further research; Elizabeth back to her undergraduate studies.

# 9 TRAPPED IN THE BLUE OCEAN

April. Five months had passed since Amik had escaped with her life from the hunting stellar sea lions. She had swum south for four months and had reached the waters off northern California. This was not as far as more mature female seals like her mother, who swam much farther south to waters off southern California, close to the Mexican border, but her swim had been a voyage of discovery. It had not been an easy journey. She had drifted south and east, well south of the Gulf of Alaska. In the open ocean she had learnt to hunt, to fish, always remembering the successful associations with food. She learned that when birds worked the water there was food; she learned the noise and call of dolphins and knew that where they were, often there was fish. And she herself listened for the songs of the fish, those underwater noises from the swim bladders that many fish emitted. By associating food with configurations of other sea creatures, or of seascapes, she rehearsed the well-known swimways used by her ancestors.

In March she had turned and begun to retrace her steps, north, back to the Bering Sea. There were five yearlings in the loose group which seemed to travel together, although there was very little interaction between the young seals. Sometimes Amik would be on her own for hours on end, then suddenly she would catch sight of another head in the waves and would know other fur seals were around, quiet shadows in the vastness of the Pacific.

Clouds of krill filled the ocean, like flakes of ice in a slow-falling snow storm, and she hung motionless amongst them. Spring had come and with its warmth a quiet stillness. Surface waters no longer mixed, in the violent turmoil of winter storms, with deeper, colder layers, and the ocean became stratified once more. Phytoplankton cells emerged from the dark depths of stormbound winter and grew rapidly. The larval copepods climbed up through this green bloom, moulted into maturation and bred. The eggs they produced hatched into tiny nauplial larvae which fed on the abundant phytoplankton. Everywhere there was life, food in abundance.

For a month or more she had noticed her cousins, the California sea lions, swimming south, moving back to their breeding colonies off Baja California. She swam on against the current of their migration. The krill clattered and chattered, treading water with rapid movements of four pairs of spindly legs; their big black eyes and dark stomachs stood out bold against the blue translucence of their bodies, which bumped and tumbled in stiff, jumbled movements about Amik's head.

The young female seal spotted the chestnut-brown coats of a group of six sea lions heading towards her and felt the scar at her side, remembered the sea lions of the Pribilofs. She and the other yearlings waited, watched. The sea lions swam by, but one strong male peeled off from the group, opening wide his mouth. Bubbles, globules of air, wobbled around the yellow canine teeth. Amik started, turned sharply to face him head to head, shoulder to shoulder. He weighed eight times as much as she did. He moved swiftly, deliberately, straight towards her, krill parting at his head, tumbling about in his wake. She snarled with gaping jaws, but he flicked, pushed on his forceful front flippers and sped up, above her; bit deep into the side of a steelhead trout and continued without pause to the surface of the ocean. There he tossed his prey low into the air, turned it, head on, into his mouth, flipped it again and then again: for a moment it fell sideways almost out of his jaws, but he bit firm. Then, with one last movement of his powerful neck, he opened his mouth wide and took the whole fish into his deep throat, fins and all, head first.

Amik turned quickly to face north and porpoised off. She swam for a kilometre before she slowed, paused and then sank below the surface. So she meandered, following the tight corridor of her species' migration route to the misty island in the fish-rich sea, far to the north; remembering, learning, programming her own programmed mind with facts about the annual return; always staying in waters no deeper than two hundred and twenty metres, at the edge of the shelf. She entered the warmth of the North Pacific Current. The temperature was too warm for her needs but it was food and the drive to swim north that kept her going, drew her irresistibly into and through these waters.

May blossomed into June. The ocean flushed green with plankton, thriving under the warm, direct rays of the overhead sun. Sizzling shrimps browsed noisily through the richness, and fish fed ravenously on the krill. Amik was nearly one year old. The instinct—her instinct—continued to drive her northwards, back to the Pribilofs. Now she swam alone, a silver head in a wide, glistening sea of summer. It was day and she was resting, her small ears erect and alert, listening. Flat across the water, above and below the surface, was the noise of engines, propellers churning. They were sounds she had heard before, at a distance; they meant little to her. She lay on the kelp mat half drowsing. The sun, in that clear sky, was low in the west; streaks and sheets of reflection ricocheted, bounced off the irregular, dark, shining surface of seaweed.

There was other flotsam in that mat: a plastic bottle, several small planks of wood. The bottle began to bob. The wood heaved. The kelp undulated. It was the approaching wash of a net being dragged, coming at her through the water. She knew nothing of nets, apart from her brief experience on the shore of St George when she was a very young pup. Then she had been interrupted in her investigations by the return of her mother. Occasionally she had seen discarded remnants floating by, as she rested or swam lazily at the surface of the sea. Always passive, those nets, they just drifted with currents and tides. This net was different, and at first she did not recognise it. It was the approaching boat that frightened her, and the white plastic floats, coming at her from all sides, puzzled her. She rolled over and sank down under the surface. The sunlight was still strong enough to illuminate the top few metres of the sea.

Below that was the darkening circle of deep water.

At first she did not see the net at all. Then it glinted. The open mesh, fifty-millimetre squares of thin, strong nylon thread, sparkled in lines, now down, now across, as the movement and the current waved through the gill net. Lazily she swam towards it, noted its form, sniffed, and felt its texture with her whiskers and nose. Bubbles streamed from her mouth and the fur immediately above her front flippers. They lingered, glinting also, and slowly floated to the surface, drifting in and out of the mesh. At first she swam with slow grace, her back flippers trailing. She rolled over, swept back on herself and noticed that the net was encroaching, creeping all round her. Suddenly she panicked and swam hard into a shadow where the sun did not pick out the line of nylon threads. It was clinging, and fish were around her, struggling too, all trapped. Salmon, threshing about with their powerful tails; big fish, small fish, any kind of fish, all about her head.

She was caught and dragged. An inexorable force was carrying her forward, pulling her, moving her in a direction she did not wish to go. She struggled, squirmed, wriggled, to no avail. The white floats about her were getting closer, the noise of the engine was very close. There was a banging of chains, a clattering of tackle, human voices, the cry of gulls hunting, diving and feeding. Now she was being lifted out of the water, up through the air. That curious sensation of the air pressing down, and yet there were no unyielding black surfaces; no rocks on which to rest her heavy body. She had felt this when she was a pup, the weight of air, but not like this. Then it was all about her and pressing down. Now the net was clinging close.

Suddenly the speed of drag increased, her back flippers were bumped roughly against a smooth, metal vertical surface. Up, across the riveted ship's side, her head in the taut net, forced backwards, outwards away from the boat itself. Her eyes caught sight of the horizon rocking up and down, side to side; the sky was down, or she was upside down? The sea was up above her head? She heard gentle waves breaking against the ship's bows. She was dragged up, three metres above the calm surface of the sea, then savagely wrenched forwards, a rasping drag across the plates of the ship's deckside, her head spinning, and the final fall.

It was a close, ringing sound, the metal deck sliding under her. She was trapped and her jaws were held firmly. Her eyes blinked hard against the tight net cord which dug into her face and pulled roughly across her left eye. For a moment she lay still. Eight men in white helmets, thick black waterproofs and stout boots, crowded in on her. Fourteen legs moved forward. She writhed again. The rubber boots clomped, echoing on the metal sheets. Littering the entire deck were piles of the wispy fine gill net and the small white plastic floats. All over the deck was the violent flapping and floundering of caught fish, salmon gasping, jumping with no hope of escape. A heavy pair of boots was at her head. A firm hand came down, pressed her chin harshly against the cold wet deck, squashed her nose, disfigured the shape of her face, pushed awkwardly at her left eye. She squirmed to bite, tried to pull at the muscles of her own face to force open her mouth. She tried to wrench her head round, but her neck was also firmly held. She writhed, wriggled, tried to bite again, brought her back flippers round, but the men were pulling at the net. And she was freed from the netting, all except her head. Hands held firmly to her back flippers. She felt them tugging gently. Voices were shouting, there was a banging and clanging, fish thundering and juddering against metal sheets.

The net was torn efficiently from her face. She tried to lunge forward with her mouth agape, snapped, but the hand moved too quickly away from her and she was pulled back, twirled round. For a moment she was floating, sensing the weight of the air and the noises, but there was no net, no clinging hold at the back of her neck. And she fell back into the sea. The deafening sound of splashing, bubbles, air in her fur, the underwater sounds of the ship's engine, and she sank two metres, turned, turned again, and shot forward, porpoising out of the waves; forward again another twelve metres, porpoising on and on, away from the boat, safe from the net.

In August she began to swim south-east. All summer she had skirted the Canadian coast and had seen the trees of Vancouver Island coming down to the sea. She swam close to that island because the continental shelf break was very close to shore and that was where the fishing was good, but in the Gulf of Alaska she had turned, veered south of Kodiak Island past Shumagin and Sanak, and then, swinging north-west, had breached the Aleutian Chain between Unalaska and Unimak. The volcanoes towered above the swirling waters of the passage; the kelp beds were dark, like a thick forest canopy, and she the bird of the water, flying, pushing through the leaves, the hanging fronds.

The cold water swirled south, and she was a yearling probing the northward passage.

There was no urgency in her journey. She was exploring, testing, tasting the route, almost meandering but always learning and remembering. From the depths of her mind and the aeons of evolved instinct, she was dragging up the northern route, the true passage for the northern fur seal. Now, from time to time, she saw more heads, fur seal heads, peeping up from beneath the kelp beds or free-swimming in the cold water. The sea around the Aleutian Chain of islands was shallow, no more than sixty metres deep, so she often reached the ocean floor in her hunting dives.

Occasionally she noticed another creature. It looked like a fox, but swam at the bottom of the sea. She paid careful attention. One day, very late in August, she had just lost sight of such a creature, darting between the kelp beds, when she surfaced and in the evening light picked out the shape of a short log, floating low in the water. It lay in the middle of a kelp bed—just over a metre long, dark brown, but one end more of a straw colour. She swam slowly towards it, and it moved, although apparently unaware of her. It was an animal, holding a shellfish in a front paw, and although Amik could not see the stone on its stomach she heard a crack as the sea otter brought down its paws with great force. The noise unsettled her, she submerged. Curiosity brought her to the surface again, and she watched as the creature nibbled and fed. Then suddenly she heard a distant sound, one which sent tremors of terror down her spine, a remembered sound. She was close by Dutch Harbour. The boats were coming out and she was in line of their route. The otter turned turtle, ears pricked, seemed to hear what the fur seal heard, and vanished from Amik's sight.

She turned in the kelp. On the line of her horizon were beads of light laced together by the bobbing shadows of the seaweed around her head. Night was approaching and the boats were coming closer, trawlers setting out to find the pollock. Away from them she swam, north and away. She had no idea where to go but she clung to the vision she held of the route she must take. She swam steadily and hard for more than an hour and she felt the strain of the journey. North she swam, into the Bering Sea teeming with krill and all the life that came to feed off the richness of the cold waters. Closer and closer to the Pribilofs.

It was not until early October that she reached St George. There was no scientist on the shore and no research assistant in the hide. Money for the research project was tight and that year general observations had been reduced because Dr Alan was reserving all his major funds for satellite tracking. So no one saw Amik come and no one saw her go in the cold, damp autumn of her first year.

# 10 PERIL AT SEA

December 25th. A warm winter evening and the soughing of a quiet-voiced wind in the deserted streets of down-town Los Angeles. Long, cool whispers of an evening breeze blew pleasantly through the orange trees, played a sleepy tune and rattled palm fronds in the grounds of low, opulent houses. Gaggles of important people sipped red wine around the garish blue waters of pure white swimming pools. On the crowded beach a calm sea sighed, and huddled groups of young people laughed, cavorted and lit fires, while gulls flew back from the open ocean to high roosts on warm cliff faces, crying shrilly through the bright white spray.

Eighty miles due east, out in the Pacific, offshore oil platforms near Santa Barbara were topped by fierce burning flames, licking the open sea breeze, flickering; ghostly night-lights, shimmering across the blackness of the slowly heaving ocean. Below the waterline there were rumblings of activity which echoed deep through the sea.

The same blue sky which darkened over Los Angeles and the oil rig also faded over quiet ripples lisping and purring around a silver head. The wide waters of the Pacific heaved a deceptive calm, the same that had so beguiled Cortez when he first set sight on the ocean four hundred and fifty years before. Cold, clear, dark waters, and in them that silver head moved quietly, disappeared, reappeared and looked round. It was a female northern fur seal, eighteen months old: Amik. She was now a strong, healthy juvenile, nearly a metre long and weighing twenty kilograms, recognisable as an individual of her species by a scar, much faded, but visible down her left side.

She began breathing deeply, gazing up at a sky bright with stars and the moon nodding down to the western horizon.

Suddenly she exhaled to limit the amount of nitrogen in her lungs, and leaped into the warm air, curving back in a continuous motion as she dived; streams of water flashed from her dripping flanks, burst and thronged the growing circle where she had sunk back into the ocean, leaving the lonely seaway quiet again, just the hushed susurrus of the sea and the mewing of the skuas.

She was hunting. Down, down she swam, her heart-beat slowing with every metre she descended. Down to thirty metres. Bubbles of air were pressed out of her heavy coat by the weight of water and went racing in silver streams past her head; her brown eyes were wide open, and in her lungs the alveoli, those minute, air-filled sacs through which gas exchange takes place, collapsed, stopped her getting 'the bends', so feared by deep sea divers. Forty metres and more she dived, still trailing behind her a stream of bubbles, phosphorescent in the dark waters.

As she swam, all her senses taut, keyed to detect any sign of prey, she felt a turbulence, the sideways whack of water. She rolled in it, bobbed like a duck in the wake of a boat. Whatever it might be, it was big and close by. Out of the dim, dark mist of the night-time sea came the growing shape—enormous, dominating, swimming past her: a blue whale. Amik's ears were full of the clacking of shrimps as the whale came by, all open mouth and bubbles. The coarse bristles lining the inner edges of its baleen plate made up an amazingly efficient filtering system. With it, the monster of the deep strained the shrimps out of the water, held them back on those bristles. In the wake of this great movement the seal rocked gently, felt a thousand bubbles rise as the largest animal that has ever lived swam by in full, overwhelming majesty, leaving her surrounded by the continuing sounds of the krill, noises more akin to the sizzling of a thousand frying eggs than the supposed silence of the underwater world.

In the darkness, the seal was aware of a dull, rising glimmer, and the cold light from tiny luminescing organisms floated and surrounded her. It was the recognition signal for the swarming instinct of tiny crustaceans, myriads of minute creatures lighting their undersides to help them blend into the faintly lit background. In a writhing, wriggling mass they floated through the sea, and following them came slightly larger predators, such as jelly-fish, and following them larger predators still, the herring—a shoal of them, weaving and wavering round and round in a tightly packed ball.

In the glimmering light the seal identified her prey, saw it dimly ahead of her. Her brown eyes were accustomed to the gloom, but she responded more to what she felt—the vibrations, the twitching of her whiskers—than to her vision, and to her ears, not flat against her sleek head but slightly proud, parallel to her fur, absorbing what tell-tale sounds there were. And there were thousands upon thousands of bubbles, glistening, wobbling, coming at her, as she swam forward with increasing speed.

The ball of fish was now directly ahead of her, a large, glistening mass suspended, twirling, in the purple water, and clouds of bubbles streaming back from the fish, flooding over the seal. She flicked her tail flippers in aim, pushed on her wide front flippers with enormous power, and sped with the noise of a wailing banshee. The ball exploded into a thousand fleeing fish as the seal suddenly tightened her neck muscles, gave a short, deliberate push to the water around her mouth, and snapped shut her gaping jaws. Fish tumbled over her head and all around her, while she flung her head from side to side, her canine teeth tight into white flesh, and sped to the surface with her capture.

She finished her meal, and in the silver cool of night rolled over, clapped her front flippers and rubbed her nose. There was light enough at the surface to see that she was alone, as she often was during her winter months. The moon was still up in the west. Pisces and Ceres sparkled crystal in the darkness. The enormous water pressure of her deep dive had squeezed the air out of her fur. That air had to be restored or she would lose body heat, so she began by scratching, running the air back into her coat, aerating her fur. Meticulously she combed every square centimetre, then slowly, leisurely, she turned, rolled, swam forward with increasing speed, porpoised and dived again.

Living in the North Pacific was not easy, although the food supply was plentiful enough. The seal had spent the last winter off the western side of the great continent, around the Tropic of

Cancer, in the cool waters of the Californian Current. Cold waters are rich waters. The density of life in the sea is bound up with the abundance and circulation of nitrates and phosphates. These two elements are needed for the growth of plant life and there tend to be more nitrate compounds in cold water than in warm. Consequently, cold currents and upwellings of bottom layers make extra supplies of nitrates available to life at the surface—hence the richness of the cold southward-creeping Californian Current off the coast of Lower California, made richer by the light and heat of the weather over that part of the ocean.

The continual growth of minute marine plants attracted a wealth of crustacea and small fish, and these animals provided food for larger creatures. Instinctively the young fur seal knew this; that was why she migrated south every year from her breeding grounds in sub-Polar waters. Any farther north and the living was not so good. Polewards of forty degrees latitude, the North Pacific Current, drifting in from the west, warmed Seattle, Vancouver Island and the southern coast of Alaska. But off southern California the ocean was that much cooler, with deep upwellings of cold Arctic waters rising slowly to the surface.

However, it was not just the presence of that cool current which provided the nutritive elements, but also the constant overturning of shallow waters at the edge of the continental shelf. Wave action and tides helped make the full resources of the whole mass available for plant growth. And there were some areas of ocean, certain localities, where the mixing of waters was most pronounced; these yielded most fish. The female seal had searched out these places, remembered them and fished them well.

She dived again, her ears picking up the sounds of a group of white-sided dolphin chattering and crackling like interference on short wave radio. Spluttering and sizzling, they swam past her. They were just under twice her size, grey with flashing white bellies. One came very close to investigate. In the gloom she saw the rounded snout and short black beak, the narrow, pale grey stripe above its eye, running the length of its torpedo-shaped body. It swam by. No threat to her, for it was not a predator of fur seals, the dolphin searched out the same herring that she had fed off so well. Like the seal, they followed undersea escarpments at upwellings of water where fields of plankton encouraged shoaling fish.

Almost lazily she flipped over, changed direction and followed the dolphins. It was likely they would work the sea together, those dolphins, herd the fish to make the hunting easier. She, Amik, was the lone opportunist and would follow them because she knew from previous experience that there would be more than enough for her too, as herring tumbled in frenzied flight away from the ravages of the sharp-toothed mammals.

Midnight. Tired after her dives for food, and groomed well, she rolled sideways in the water, lifted her two rear flippers in what at first seemed a very awkward movement, and grasped her left front flipper, holding it tightly so that three of her four flippers were high out of the water in an arch, almost like an elongated jug floating half submerged, a dark smudge on a dark ocean. Her flippers had no protective fur covering, and although the Pacific was not half as cold as the Bering Sea where the mature male seals were spending the winter, she still wanted to conserve heat, so she rested with three flippers out of the water and just one in the cool sea to stabilise her sleeping form.

* * *

Two years passed by. Amik followed her annual cycle of migration, swimming up to St George, but unlike the male juveniles she did not pull herself out of the sea and onto the hauling grounds; instead she swam offshore. From her position there, at the very edge of the rookery, she could watch the mature males and females occupy territories, quarrel, fight and roar. During those two seasons at St George she also learned about the feeding grounds. She left the island and swam off to hunt in the richness of the Bering Sea, and often, on her return, she tested her knowledge of the very edge of the land and coasted in on a large wave, drifted very close to the rocks of grid square 14 and was able to identify, near a large black rock, eight or nine females crowded together.

So now Amik was in her fourth season. It was early April and she was a fully mature female. Having overwintered off southern California she was beginning to move north again, through the blue waters of the open Pacific. Below her the edge of the continental shelf dropped sharply away to depths below 600 metres. This year she travelled with more purpose: the beginning of her migration journey seemed to have more urgency about it, and she would arrive at St George earlier than she ever had before. This year she would climb up onto the rookery itself and try to find a spot for herself as close as she could to that black rock on grid square 14.

A leatherback turtle swam by, diagonally across her own migration path. It, too, was migrating. The fur seal did not see it in the darkness, nor did she notice two triangles, cutting cleanly through the ocean, one hundred and fifty metres away from her. They were about three metres apart and stuck up half a metre out of the water. They were moving slowly and smoothly left to right in a clockwise direction, describing a circle with a circumference of a kilometre, with Amik in the centre. A radius of just over one hundred and fifty metres separated her from the fins: unknowingly she was trapped by a great white shark.

A long, pointed snout with a crescent-shaped mouth on its lower surface held strong serrated teeth. The shark pushed powerfully through the water, more than six times the length of the seal and four times as big as the turtle; grey on top and with white underparts, the top lobe of its tail was slightly longer than the lower and it was that top lobe which whispered through the calm waters. The creature had no need to surface to breathe: behind its head, in front of the first dorsal fin, there were five gill slits on each side of its body. It breathed like a fish and its skin was rough and prickly, covered in tooth-like scales. It was a very large and extremely aggressive carnivorous fish and it ate just about anything, dead or alive.

Quiet waves lapped gently at the slow-swimming head of the fur seal. She blew slightly, the sound reflecting off the undulating surface waters. The leatherback turtle, on a different radius of the shark's hunting circle, flapped forwards in a south-easterly direction. Its long fore flippers, with a span of about two and a half metres, were propelling it towards a floating jelly-fish, its weak scissor jaws preparing for the meal.

Although it was a turtle, it had no horny shield on its shell, no scales and no claws. Its carapace resembled hard rubber. The fur seal swam as if she was at one o'clock on that hunting circle, the leatherback as if it was at seven o'clock and the shark at midnight.

Ahead of her the fur seal sensed movement. Prey, perhaps. She swam faster. The shark continued its slow circle, moving towards a shoal of anchovy. The fur seal veered slightly to her right, following the approaching line of the shark, and suddenly she picked up the bubbles and clatter of the anchovy. The shark zoomed forward. The leatherback continued to head for the jelly-fish, and the seal moved forward, too.

The shark sliced straight through the anchovy, snaffling like a goldfish at tadpoles. Almost too late, the seal saw the shadowy whiteness of the great predator; she stalled with the perfect precision of her left front flipper, turned due north, sped to the surface and porpoised away as fast as she could, the anchovy tumbling and swirling in a noisy, glistening cloud, while here and there small specks of blood dispersed rapidly in the maelstrom, as the shark turned and turned again, eating voraciously.

In the Bering Sea on that same night, a ten-year-old male fur seal was feeding, too. Late the previous season, Amik had seen him occupy most of grid square 14 and the adjacent parts of grid square 13 which comprised his territory. He was a very strong bull and had extended his territory, through sheer fighting prowess, over the best land close to the breakers but not flooded by the tide; during the coming season he would expect to mate with more than forty females. If Amik could struggle back he would mate with her.

Already the waters of the Arctic were beginning to bloom. The early April sun had brought the first flowering of plankton in the sea and on the pack ice which glittered gold with diatom algae. In the open sea the early predators were about. Delicate-looking comb jellies, hardly more than mobile digestive tracts resembling tiny translucent barrage balloons, devoured vast quantities of minute crustaceans. Feeding on the comb jellies and other small organisms came shoals of capelin. These small salmon, barely twenty centimetres long, were very effective converters of plankton into food for other higher creatures.

The bull seal fed as well as the shark, three thousand kilometres to the south—as well he should, for he had to build up his supplies of fat quickly in preparation for the long summer fast. Soon he would be battling again for space on the cold northern shore of St George.

The march of the seasons continued . . . April . . . May . . . June . . . The annual rise to the surface of fish and squid which had been driven into deeper water by the severity of winter, coincided exactly with peak pupping time on the seal rookeries. By the time Amik reached the waters of the Pribilofs it was early August and the food supply in the ocean was at its greatest.

There were six rookeries on St George—two on the south side, not far from the new harbour works, and four on the north. To reach her own place Amik had to swim past Staraya Rookery and North Rookery. Morning dawned as she approached High Bluff, and ahead of her she picked out the floating, bobbing shadows of puffin feeding well. She swam on through the kelp beds. A few seals came swimming past her, making their way from Staraya out to the hunting grounds. Amik saw them and ignored them; she was getting used to the ghostly white of the sea floor where columnar basalt, covered in coralline algae and browsing urchins, looked like some gigantic abstract mosaic, and crabs wandered, carrying a disguise of seaweed in an attempt to escape predation. Starfish searched out mussels, forcing open their prey with skill and strength. At depths below twenty metres Arctic scallops browsed, filtering food from the swirling water to build up their reserves of starch ready for the lean times of winter.

On she swam, past the bay which held both North Rookery and the small harbour around which huddled the weather-boarded cottages of St George. On past the skin plant. Sea Lion Hill and Seal Lake lay inland from her; she would never see them, just as the islanders would never see the underwater landscape of her own environment. She pushed hard with her front flippers, close now to the shore and East Reef. Ahead of her was the hundred-metre band of marauding male juveniles, all waiting for her. She sat up high in the water, saw them all, bobbing, weaving and porpoising, then she dived, swam fast and straight. A wave broke noisily over her head and she was rushed onto shore, close to grid square 14. There she climbed immediately up onto the black rocks and flopped down heavily, panting and feeling the pressure of the open air.

She never mated with the strong male. By the time she arrived at the rookery, a new male had replaced the bull of the season. However, it was quite close to a large black rock, within sight of a curiously familiar ten-year-old female, that she did mate with a young male fur seal; so she carried away the seeds for delayed implantation, south again across those thousands of kilometres of wide ocean, back into the North Pacific off California.

* * *

On March 24th, in the fifth spring of her life, Amik was off Monterey, five hundred kilometres north of Los Angeles; following an underwater canyon, she had come up close to a set of offshore oil rigs. It was dark and she was busy, hunting. The flames and lights from one oil platform flickered and burned, seared across the still waters of the ocean. Here and there the rainbow colours of dispersing diesel oil caught in the gleam of reflected lights. The oil was just a tiny amount of effluent from shipping which travelled to and from the platform to the mainland. Occasionally Amik tasted that foul water, did not like it close to her fur.

Meanwhile, more than three thousand kilometres to the north, just off the coast of Alaska in Prince William Sound, Captain Joseph Hazelwood of the oil tanker *Exon Valdez* was about to make an historic radio call. The ship's third mate said he had hit a reef after veering to avoid ice. The Captain called the US Coastguard over the radio and, in a droll, laconic voice muttered: 'Evidently we're losing oil and we're going to be here a while.' A quarter of a million barrels of crude oil were spilled that night.

Completely unaware of the catastrophe, Amik swam north again. This time she was pregnant, carrying her first pup deep inside her. Within a month of the spillage, hundreds of unemployed Alaskan workers had been issued with black, red or fluorescent orange waterproofs. They sat on blackened beaches being paid twenty-five dollars an hour to put oil-killed carcasses into polythene bags or scrub rocks clean of oil, one by one. Five hundred and sixty kilometres of beaches had been affected before the late spring storms pushed the oil slicks out of the Sound and into the open waters of the Gulf of Alaska, only three hundred and twenty kilometres north of Amik's migratory path. The men in fluorescent suits had half-cleaned 800 metres of beach.

Amik veered west three hundred and twenty kilometres south of the slick which reached out along the shores of the Gulf like a great slug, leaving a dark, slimy trail, a killer glob. By the time the young pregnant female had reached Kodiak Island the slick was curling round the Kenai Peninsula and seeping into Cook Inlet. The fishing towns of Seldovia and Homer, on either side of Kachemak Bay, were trying to protect their communities with makeshift log booms. But every time a storm blew out of the north, south or west, another beach or fishing community was stricken.

Doctor Steinberg had rushed up early to St George, by way of Juneau, the capital of Alaska, and Valdez, the site of the oil terminal and centre for the clean-up operations. He was anxious to find out as much as he could about the effect the spillage would have on the fur seals. He had surprised Alex, when he radioed ahead to announce his arrival.

'Come round for a meal, bring your father.'

It was a very sober evening. The decline of the fur seal, and indeed the decline of most of the animal life in the North Pacific, was in their minds, and this latest catastrophe, the oil spill, seemed to add to the worry, almost made their conversation despairing.

'It's as if we've actually killed the land,' said Alex.

Alan looked puzzled.

'Well, you know,' continued Alex. 'When you walk on a beach when the tide is out, there's a sort of squirting—you know—the animals underneath, worms wriggling round, starfish, clams. Well now, according to the fishermen and those guys from over on the mainland, it isn't like that any more. Not on the beaches. That oil's trashing everything.'

Alan thought about it. There was a long, long silence. Suddenly Alex broke in.

'The harbour's nearly finished. But the island council are having problems getting the army to dredge it clean. Apparently they've found high levels of mercury in the mud, and they're frightened that if they disturb it, it'll pollute the sea round Zapadni.'

'I've not heard about that,' said Alan.

'Reckon it might kill off the seals.'

'Is that true?' asked Alan in disbelief.

Alex nodded.

'God, they're neurotic these days, frightened of criticism. I'll write to them,' offered Alan. 'The mercury's a natural deposit. The seals have swum through those waters for millions of years. The storms churn it all up anyway.' The news seemed to create more anger in Dr Steinberg than in the two islanders. They sat silently, listening to his reaction. He paused for a moment, then said,

'I'm cynical. I think it all makes work for the bureaucrats. And they never make the right decisions, never actually stop the powerful lobbyists from doing anything, no matter how damaging. This oil spillage is an example of that. Everybody knew that there was nothing they could do if there was a leak. Still let them drill for oil. But a bit of mercury and . . .'

'It's politics,' grunted Vladimar. 'If St George has a harbour, what does that do for Dutch Harbour? There's only so much work. And the Aleuts, how many votes are they worth, eh?'

'Poor buggers,' murmured Alex, his thoughts elsewhere.

'Who?' asked Alan. Both father and scientist were confused.

'The seals. If they escape the killer whales and the white shark and the hookworm and the floating bits of netting and the trawlers and the gill nets, they've still got that goddam awful slick.'

Alan said nothing.

'How bad is it, Dr Steinberg?' asked Vladimar.

Alan shrugged his shoulders. 'Don't know yet. It shouldn't really affect them. Unless the winds take the oil right out past Kodiak and across their migratory route.'

'Different story for the sea otters. Have you heard?' asked Alex.

'Yes,' sighed Alan.

'They were only just beginning to make a come-back. Isn't that right, doctor?'

'Yes,' replied Alan, almost as if he was just waking up from a dream.

'They were protected in the 1911 fur seal treaty. That's how they started to do better, isn't it?'

'No doubt about it, Vladimar, your son knows his wildlife.'

'Have to,' responded Alex. 'All these tourists. Got to try to know as much as, if not more than them. Otherwise they won't like paying us.'

'No, you're right,' agreed Alan, referring to the sea otters. 'They'd gotten up to about twelve thousand, but now the oil's wiped them out in the Sound, Prince William. That's three thousand gone.'

'Haven't got blubber, you see, Dad,' interrupted Alex, wanting to finish off his own story. 'Depend on their fur entirely.'

'Mind you, in the Sound,' Alan informed them, 'the seals and the sea lions are diving in it, ingesting it, breathing its fumes, and they'll be pupping in it, too. Hydrocarbon poisoning.'

'It's the greatest environmental disaster in history, that's what somebody said on the TV.'

Alan looked at the young Aleut.

'So what are we going to do this year, Doctor Alan?' Alex asked.

'Same as always. But if you see any seals with so much as a blob of oil, I need to know.'

Alex nodded.

'I don't expect any fur seals will swim through it,' Alan went on.

'No,' agreed Alex. 'The fishermen say it's staying close to the mainland.'

'Yeh. They'd have to be pretty strong winds to bring the oil out to the seals.'

'That's what I mean,' explained Alex.

'Even so, there's no telling. We've got to be alert. It'll be a bit like net entanglement. If a few get oiled up we can clean 'em down. But most'll just die out there, in the open ocean.'

Vladimar smiled a long, laconic smile, full of critical sarcasm.

'Once upon a time—' he began. 'In fact it wasn't too long ago, a month last Friday, to be precise, just at the time when man was coming to the later stages of life on earth—a great people rose up. They were called "us", or as the rest of the world put it, the United States of America. Now this nation was proud and its pride depended upon every voter having the right to drive a vehicle as far and as often as he wanted and as cheaply as possible. "Voter" was the word most politicians used to describe people. Politicians were the witch-doctors of power.'

Dr Alan Steinberg looked at Vladimar. So this was where Alex got his storytelling.

Vladimar continued. 'And this nation discovered great quantities of raw vehicle fuel under the shores of the northern extremity of that nation. And this northern extremity was a very lonely and beautiful place. Each year many of the planet's sea animals and birds returned to breed.

'Now although most people, including the politicians, feared an evil giant called Accident who could unleash his terrifying dragon called Catastrophe, everyone thought how good it would be to pump the vehicle fuel from under the northern shore and to put it in pipes down to the nearest all-year port and load it into enormous boats to carry it to the centre of the nation of vehicles—"us", in other words. There were many learned men called scientists and protectionists, and many learned groups of people called committees, who sat and talked for hours and hours and hours. They said how delicate and beautiful the northern shore was, and how narrow the sea inlets, and how dangerous it would be if the giant Accident ever escaped and unleashed his dragon.

'So they put a charm on the giant, called Contingency Plans, and filed this charm in all sorts of different places, like a small town called Juneau, and a big town called Washington and a tiny, tiny town called Valdez. And that was that. They all went home and drove their vehicles. For the waxing and waning of two hundred moons the giant stayed charmed and asleep, and then in the spring of a dreadful year he woke up, got drunk and unleashed the dragon. End of story.'

'Not up to your usual standard, Dad,' commented Alex, draining his beer.

Alan said nothing.

# 11 ATHOW AND HER RETURN

She had been here before. The rumbling sounds, she had not heard them for eight months, while she had been alone in the warm blue ocean; now the sea was grey-green and cold, and the low sky clouded. But there was the rumble: waves against the land. And the smell, the scent of return in the water, this water.

The sea heaved and swelled. There was a translucent flatness to the surface above her, no evidence of the great colliding waters, the cold Arctic Ocean against the vast power of the Pacific. Here she swam through the enormous upthrustings, currents bringing the ocean edibles constantly to the surface. In summer, when the ice of the north retreated, the sea was unlocked, and she had returned to where plankton thrived and shrimp grew fat. And when the shrimp grew fat the salmon fed on them, and the pollock and the herring and the halibut. And when the fish were fat and plentiful, walruses, whales and seals could feed well. Nature threw this abundance of seafood to the surface, in summer, in the Bering Sea. And she had come to feed, to gather the strength for giving birth and rearing her pup.

The bright ripples, ceiling of the sea above her, shone deep into her dark brown eyes. She flicked hard with her tail and soared upwards through the cold green; broke surface, out into the greyness of the Bering Sea, and heard the roar of the land: waves on the rocks, migrating seabirds—murres with eggs, puffins coming out of their burrows, red-legged kittiwakes arguing over cliff space—birds of the open sea come to this isolated island to breed in the comparative safety of the towering cliffs. And round her ears the open air, bubbles racing, wavelets ringing hard against her head. Often each day, in the wide waters of the blue Pacific, she had heard that noise—the noise of the surface and the lone cry of the petrel and the calling of the travelling gull—but this cacophony never, not in the blue shining warmth. Only here: ten million birds fighting, mating, breeding at the edge of deep, rich waters.

They towered above her, the dark cliffs of St George. High Bluff, 200 metres straight out of the sea. For a moment the black shadow overwhelmed her and she turned her back on the land, turned again to the open greyness of the ocean.

On the island, along the beach, stood the motorboats, and on the harbour wall lay the nets and against the walls of the weatherboarded cottages, lines, harpoons, and the long clubs for knocking seals on the head. And in the cottages were television sets full of dark pictures about oil on the beaches of Alaska. So close was she to the sea, she did not see this. Not the cottages, not the new harbour, not the new runway—not even the church with its cupola of copper. But in the sky above her she saw the clouds of least auklets, high and crowded together in their thousands, great swirling masses returning to their roosts on the land after a day at sea foraging. Like bees to the hive they swarmed in the sky above her, preparing for the downward swoop into cold darkness.

Amid the noise there was a quiet cough. She did not hear it, but the mangy fox scrambling over the steep grass edge had paused to watch her. It looked over the rocks, searched for the early egg, the unsuspecting nesting bird, then coughed again, turned and hurried on through the murky gloom. The first cloud of least auklets suddenly plummeted from the sky, like floating specks of dust sucked quickly and efficiently into a vacuum cleaner. They disappeared without trace into the black shadows of the island.

Then suddenly it was quiet—the noise of ten million birds silenced by the dip of her head below the waves. The bubbles raced, streaming past. Down she swam into the cold, clear water. There was the rumble again, the loud clicking rattle of shrimps feeding and jostling in their thousands, and the sea, the waves against the land, the scent of return.

The least auklets settled down into the short night. The prowling fox returned to her den in the abandoned oil drum. And the female fur seal rolled over, arched her back in the heaving ocean and sank leisurely past the long fronds of kelp, letting the slow current take her gently away from the shore.

It was a summer evening on the island and the grass growing on the low cliff around the skin plant was fresh green. The abandoned buildings, where slaves had once skinned fur seals for the backs of rich women, were still in good condition, though empty and unused. Not that she knew that. If she had seen the low, flat silhouette she would have interpreted it as nothing more than part of the known shape of the island, and swum on. But she was deep, two score metres down, and evening or morning meant nothing, down there in the green gloom.

In many ways, because winter always clung so long to frost-bitten St George, it was still spring. In that short new grass an Arctic bunting chick was cheeping, demanding attention. And in the oil drum four young fox cubs yelped and the metallic echo reverberated all around the desolate buildings of the skin plant.

The Russians had brought the Aleut slaves to the island. The Americans had bought St George from the Russians and built the skin plant. The Aleuts had abandoned the oil drums. But it was the seals who had brought them all here. The seals had always come, every summer, to this island; and now it was evening and Amik, too, had come back.

Again she raced to the surface, streaming bubbles from her pelt. Again broke into the cacophony. She paused, turned over, lay back on the cold, unreflecting surface and rested. One flipper idly folded up onto her side, out of the water, the other down below the surface, automatically readjusting her position. She yawned, her teeth showing yellow-white. The air was tumultuous with familiar sounds, unheard for eight months or more, and she had to adjust her ears.

Suddenly she felt a close rush of water, the wake of another body swimming at her, a black shadow at her side. One, two, three—ten of them. Surrounded! She turned, porpoised, as fast as she could. But the water was live with them, all wanting her, demanding her, and she heavy with her pup. Athow (the Aleut for male fur seal), hundreds of juveniles, not mature enough or strong enough to hold territory onshore, but old enough to want. At the surface they thrashed their flippers and jostled; underwater they crowded her. The waters were full of their tail-turnings, their downward swoops, their upward lungings, coming close. Too close! Wide brown eyes staring, whiskers bristling, mouths gaping, teeth bared.

She was close to her beach. These males lay between her and the place of her birth; athow, learning, exploring, finding out about returning females, learning to identify which were the spaced landing places on the beaches where the females went, the best places for the best athow. Her land was the best land, the land of the strongest seal. Many females crossed his land because it was close to the breakers; some stayed, some escaped from him. Many strong, healthy pups had been born there. And she would give birth to her firstborn on that land, the land of her birth.

She swam at the surface, letting the offshore currents carry her towards the rocks. Above the roar of the sea she heard their calls, a hundred males calling her; but none of the calls interested her, not even the call of the strong seal, the seal who occupied her land.

The scientist sat in his hide, with his notebooks, pen and binoculars, waiting in the gloom of the hut. It was cold, but his blue eyes were sharp and alert, adjusted well to the fading light, and he saw the female seal in the sea, coming close to the water's edge. The scientist had heard what the female had not, penetrating the flimsy boards of the small wooden hut: the sounds of a score of newborn pups.

And the pup, deep inside the young female, stirred. Her own, unborn pup moved, unseen, unnoticed except by Amik herself. By now the male seals onshore had noticed her, a silver head in the breakers. They roared in defiance, one against the other. Across the length of the rookery angry, anxious eyes darted looks of fierce possession. Now those males who had not seen her in the waves were alerted. Necks stretched out, high heads on powerful shoulders. Brown manes shuddered with the sudden movements. Up they mounted onto their strong front flippers. Moved quickly about their territory. Savaged the air with their aggressive snarls of anticipation.

One came down the beach, three yards too far, pebbles rolling noisily as he carved his way of challenge. The strong male saw him coming and turned with the enormous weight of his tail. A rock went spinning across the shoreline, rattling down into a crevice, close to the sleeping head of a day-old pup sheltering from the cruel Arctic wind. The enormous bull roared, spat urgent, vengeful defiance. The challenger snarled back, made a lunge with bared teeth. It was parried. The challenger collided head to head. A jarring of bones, and it was over. In a split second the challenger knew that now was not the time to inherit the best territory on East Reef Rookery. Without turning round, he hurriedly backed off up the beach, returning to his rightful position, away from the waves and the landing places of all females.

The fight seemed to have drained the bull's confidence from his huge body with its large muscles and heavy shoulders; the certainty and arrogance seemed to have vanished. He was an old male, this was his second season here—third if you counted the few days he had spent on the shore when, as a strong young juvenile, he had forced himself onto this territory at the end of the summer of Amik's first season as a mature female. She had arrived late, too. They had not mated then, nor last year, but now she carried a pup and the fur seal cycle of life would continue inexorably. This time it was likely that they would mate, for she was arriving at the peak time of the season. Already the old bull was worn by the fasting, the fighting and the waiting. The cruel biting wind of late winter, the snow, the long, deep dives in the blackness of the Bering Sea night—all had taken their toll. He had won the fight, it was still his land. The blood trickled down his shoulder. The razor-sharp teeth of the challenger had ripped his coat. There was a twelve-centimetre long gash low on his belly, through the thick fur, and it was not the only wound. There were three more, all around his front flippers, cut when opponents had tried to cripple him. None of them had healed properly. He shook himself, and as lightly as in his first days of tenure, he slid up onto a rock very close to the sea's edge. The wind and rain lashed him but he seemed not to notice the cruel weather. He just looked out, down into the waves. The female was nowhere to be seen, so he slid back down, back to the females already on his land. Seven of them. He checked them. They snarled back at him.

There were so many seals in the water, swimming in and out, weaving around her while she dodged their attentions. She porpoised, dipped her head, felt the whack of a wave shove her against the hard, steep, island edge. For a single moment the sea was empty of the other creatures, only the hanging fronds and the swell. It lifted her, carried her to where the scientist had painted yellow lines across her rocks, and she was out of the waves.

The weight of her own body, the weight of air, unsteadied her. She fumbled about on flippers and belly, awkward on the slippery rocks, the first unyielding surface she had experienced for eight months. There was a roar of wind and rain and male fur seals, but she ignored it all. She was concentrating on the short distance into the rookery. A rattle of rocks to her right made her glance up: a large male was bearing down upon her with great ferocity, his snarling, open mouth ready to seize her, pull her to his own. She was tired with the weight of her foetus, the unnatural surfaces, the confusing noises, the pressure; she veered to the left and it was then that she saw him, the strong old seal. He had lunged across the beach and was seeing off the other male. Another furious snarl of sharp incisors, more male blood, and she was back, safe, on land.

The bull seal thudded down to greet her. He leaned forward, snarled, smelt her breath. She snapped back. She was not ready for such things and she moved onwards towards the line of seven females standing in the cold evening, watching her. Their long sinuous necks reached out and down to her. Challenged her. Perhaps some she knew from another year. She had come back to the noise of the rookery; to the attention, the closeness of other animals—so unlike the rest of her life. She was a fur seal, a female come back to give birth, not to mate with any particular male. She was not any bull seal's property by right, for he only shared the place of her birth. She would stay here to give birth because it was here, within ten metres of that very spot, that she had been born five years previously. She rested, accepting the unwelcoming attentions of the other females, all readjusting their tight positions on the crowded rocks.

For a brief moment the sun dipped below the horizon, day ended. She snarled weakly at the other females, saw the male lie down in the shadows, his head on a rock facing up beach, for even when resting he guarded his land. Then she turned slowly on her tail, circled, curled up and closed her wet eyes, resting, waiting for the dawn.

* * *

Already it was light. Streams of seabirds were flying past the rookery, five hundred kilometres off shore, two metres above the waves, all heading out to fish. Round the head of the bay, just beyond East Reef Rookery, a stellar sea lion was in the water thrashing its head about, killing a fish and fighting off the attentions of three glaucous gulls, all wanting a share in the meal.

Amik stirred. As she blinked, the red light behind her closed lids slowly gave way to the dull grey shapes of the beach. Her deep, slow breathing broke up into shallow, regular breaths of activity. Suddenly a sharp flipper bone poked her back. A noisy flurry of small stones, a deafening roar, and the heavy pounding of the bull into the waves. Another female was landing in the early light. He moved so quickly across the rocks, faster than a man. Old and tired he might be, but he could still gallop across a rocky beach, tear at the flanks of a rival, grab at the neck of a newly arrived female.

The new female floundered up onto a rock. She was climbing onto a boundary area. The bull on grid square 14 lunged, caught her deep in the thick blubber of her neck, adjusted his hold and yanked her back onto his land. The young female squealed, turned, bit out and snarled—not at the bull but at another enormous male from the adjacent land, who had grabbed her right flipper. The two males pulled, teeth in her neck, teeth in her flipper. The female writhed convulsively. The neighbour tried to readjust his hold, and let go of the flipper to grab hold of the female's torso. Instinctively the female flinched, writhed and rolled away as the defending bull dropped her and buried his teeth in the soft flesh just above the challenger's right flipper, where it really hurt, and used his shoulder to push the intruder sideways down into a crevice. Then he turned, snarled, roared and shook his head. He was ready for a second flurry of possessiveness, but the intruder was struggling awkwardly away. The old bull was still master of his own land and the newly arrived female, bruised, bleeding and shaken, was struggling over the rocks towards Amik and the other seven females, while the old bull seal turned and juddered across his rocks to taste her breath. The two creatures exchanged snarls and then the new female joined the crowded wait.

Seven females turned their sinuous necks and glowered deeply at the new arrival, but Amik had slid down a large boulder onto a smaller rock and was engulfed in convulsive jerks. She turned in circles, nuzzled her tail and a pale blue sac appeared from her vaginal opening. There was a gush of liquid staining the rock, more jerks and turns, and within thirty seconds her pup was born. A small blue elastic ball, bursting black out of the placenta, staggering at once to his unsteady flippers, a dripping, shining black pup. Her firstborn.

# EPILOGUE

The oil from *Exon Valdez* never did drift out far enough to pollute Amik's swimways, although some seals were caught up from time to time in minor oil pollution incidents with fishing trawlers coming much closer to the Pribilofs than they had in former years.

Amik herself returned to St George another ten times, back to that bleak northern shore. Each time she reared a pup.

In her sixteenth year she failed to return. Whether she was caught in a gill net, succumbed to disease and starvation, or was taken by a pack of hunting killer whales, we shall never know.